The 100 Greatest Lies in Physics

By

Ray Fleming

This publication was published, written, edited, and illustrated by Ray Fleming

Copyright © 2017 and 2020 by Ray Fleming

All rights reserved. No part of this publication may be reproduced, or transmitted, in any form or by any means, electronic, mechanical, photocopying, recording, or otherwise, without the prior written permission of Ray Fleming.

Published in the United States of America

ISBN-10: 1544721803
ISBN-13: 978-1544721804

Acknowledgement

A big thanks to my sister Elizabeth and brother in-law Glenn for letting me stay at their house for a few months while I finished this book, and my nephews Colin and Connor who tolerated my presence.

CONTENTS

The Trouble with Physics	1
Lie #1: There is no Aether	8
Lie #2: Michelson & Morley Disproved Aether	14
Lie #3: Wave-Particle Duality	19
Lie #4: Photons are Elementary	22
Lie #5: Virtual Photons	28
Lie #6: Aether is Virtual Photons	31
Lie #7: Photons are a force carrier	34
Lie #8: Electromagnetic Fields are Not Real	36
Lie #9: Electromagnetic Theory Explains Motion	40
Lie #10: Magnetic Monopoles	44
Lie #11: The Speed of Light is Constant in all Reference Frames	46
Lie #12: The Speed of Light is Constant for all Observers	50
Lie #13: Special Relativity	53
Lie #14: Length Contraction	56
Lie #15: Space Contraction	60
Lie #16: Time Dilation of Space	63
Lie #17: Action at a Distance	65
Lie #18: Gauge Bosons	67
Lie #19: Mass is Intrinsic	71
Lie #20: Point Mass	75
Lie #21: Matter and Antimatter are Intrinsic	77
Lie #22: Inertia is Intrinsic	81
Lie #23: There is no Mechanical Force	84
Lie #24: Physics Explains Mechanical Motion	88
Lie #25: Dark Energy is not a Force	91
Lie #26: Gravity is a Fundamental Force	93
Lie #27: Dark Energy Does Not Cause Expansion	96
Lie #28: Non-Space	97
Lie #29: The Big Bang Can Ignore Dark Energy	99
Lie #30: The Big Bang Can Exceed the Speed of Light	101
Lie #31: The Inflation Hypothesis	104
Lie #32: The Cosmic Microwave Background Proves the Big Bang	106
Lie #33: The Big Bang Can Violate Conservation of Energy	110

Lie #34: The Big Bang	113
Lie #35: Cyclical Universe	115
Lie #36: The Speed of Gravity is the Speed of Light	117
Lie #37: The Speed of Electromagnetic Fields is Infinite	120
Lie #38: Electromagnetic Fields Propagate at the Speed of Light	123
Lie #39: The Horizon Problem	125
Lie #40: General Relativity Explains Gravitational Acceleration	126
Lie #41: Van der Waals Forces do not Explain Gravity	130
Lie #42: General Relativity can Ignore Vacuum Energy	134
Lie #43: Curved Space	136
Lie #44: General Relativity	138
Lie #45: Gravitational Time Dilation	140
Lie #46: Gravitational Time Dilation is Proof of General Relativity	143
Lie #47: The Equivalence Principle Proves General Relativity	144
Lie #48: Dark Matter	146
Lie #49: Gravitons	148
Lie #50: Gravity is due to All Forms of Energy	149
Lie #51: Gravity is Due to Mass	151
Lie #52: The Pauli Exclusion Principle	154
Lie #53: Charge is Intrinsic	156
Lie #54: Point Particles	159
Lie #55: Small Electrons	161
Lie #56: There is No Repulsive Force Between Electrons and Protons	164
Lie #57: The Schrödinger Equation Describes Electron Motion	166
Lie #58: Physics Explains Photon Production	171
Lie #59: The Neutral Pion Quark Model	174
Lie #60: The Other Irrational Meson Models	179
Lie #61: Quarks	182
Lie #62: Muons are Elementary Particles	184
Lie #63: Tau Particles are Elementary	186

Lie #64: Electron Neutrinos are Elementary	188
Lie #65: Mu & Tau Neutrinos are Elementary	193
Lie #66: Neutrinos Have Mass	197
Lie #67: W & Z Bosons Mediate Weak Interactions	198
Lie #68: W & Z Particles	200
Lie #69: The Strong Nuclear Force is Due to Gluons	202
Lie #70: Gluons	205
Lie #71: Electrons do Not Feel the Strong Force	206
Lie #72: Leptons	209
Lie #73: Protons are Not Elementary Particles	210
Lie #74: Neutrons are Not an Electron and a Proton	213
Lie #75: Particles Have Relativistic Mass	216
Lie #76: Mass is Due to the Higgs Field	218
Lie #77: The Higgs Boson	220
Lie #78: Bosons are Elementary Particles	222
Lie #79: The Table of Elementary Particles	224
Lie #80: The Copenhagen Interpretation	226
Lie #81: Spooky Action at a Distance	229
Lie #82: The Many-Worlds Interpretation	231
Lie #83: The Wave Model of Quantum Mechanics	235
Lie #84: The Kaluza-Klein Theory	237
Lie #85: String Theory	239
Lie #86: Extra Physical Dimensions	242
Lie #87: The Speed of Light is Fundamental	243
Lie #88: Faster than Light Travel	248
Lie #89: Tachyons	250
Lie #90: The Remaining Hypothetical Particles	251
Lie #91: G is a Fundamental Constant	252
Lie #92: The Inverse Square Law for Gravity	254
Lie #93: Degeneracy Pressure	246
Lie #94: The Neutron Star Size Limit	260
Lie #95: Relativistic Black Holes	263
Lie #96: Hawking Radiation	266
Lie #97: Singularities	268
Lie #98: Wormholes	269
Lie #99: Matter Production Violates the Principle of Conservation of Energy	270
Lie #100: There are Four Fundamental Forces	273
What Now?	277
Dishonorable Mentions	280

The Trouble with Physics

> *I feel very strongly that the stage physics has reached at the present day is not the final stage. It is just one stage in the evolution of our picture of nature, and we should expect this process of evolution to continue in the future, as biological evolution continues into the future. The present stage of physical theory is merely a steppingstone toward the better stages we shall have in the future. One can be quite sure there will be better stages simply because of the difficulties that occur in the physics of today.*[1]
>
> Paul Dirac, 1963

This book is not a part of the war on science, but rather the battle for the integrity of science. Physicists have come to believe a conglomeration of theories that are largely inconsistent with each other, and more importantly, inconsistent with the physical evidence. Dirac's dream of a much simpler physics has not been realized. If anything, physics has become increasingly complicated. Physicists have imagined that ultimately, we will describe the universe with a single force, a few elementary particles and a few fundamental constants, but nobody wants to do the hard work, to cut away the dead flesh of bad physics, until we are left with the simplest of all possible theories.

There is a complete absence of critical thinking displayed in mainstream physics. Nowhere are those inconsistencies greater than with theories that fail to account for the energy of the quantum vacuum, the zero-point energy. Most of the science of the 20th century was developed while ignoring zero-point energy, or even by selecting against theories that were inclusive of zero-point energy. The vast majority of physicists now recognize that zero-point energy truly does exist and yet they fail to reassess the theories put forth by scientists who rejected the existence of zero-point energy.

Other important theories ignore basic principles of physics, such as the principle of conservation of energy or the speed of light limit for the movement of bodies of matter. The most popular hypothesis that ignores both of those principles is the big bang model. There are many other theories that ignore these and other important physics principles. The selective application of sound physics principles shows a lack of critical thinking on the part of certain physicists, encroaching on professional malpractice.

On the other hand, physicists have invented rules that appear to work to describe what we see and have dubbed these rules principles, when it turns out that they are not grounded in a true physical understanding of our universe. Particles do not 'know' the rules; they 'know' their physical limitations. Some of these made up rules are treated with more respect than real principles which are well grounded in physical facts.

Errors in physics theories then infect other physical sciences such as chemistry, astronomy, geology, and biology. Physics is the simplest of all the physical sciences while at the same time being extremely complex. It is, or at least should be simplest in terms of its fundamentals—the elementary particles and fundamental forces—but as those fundamentals are applied to more and more objects and interactions, it becomes exceedingly complex. The other physical sciences add to that complexity even more, so troubles with fundamental physical theories are magnified. An error in the understanding of a fundamental force such as gravity renders numerous astronomical theories invalid. Fortunately, observation is still important to science, as the results of well-conducted experiments and observations will outlast bad theories.

Gravity is not brought up here incidentally, because if one ignores zero-point energy, there is not enough mass for the general relativity model to be correct, and if one does not ignore zero-point energy, there is way too much energy for it to be correct. These are the types of fundamental failures in critical thinking that permeate the standard model in physics.

On the other hand, the most fundamental non-physical science, mathematics, is constantly in a battle for the attention of physicists, as mathematicians and mathematical physicists are always seeking to find a physical phenomenon that fits their pet mathematical theory. Mathematicians and mathematical physicists want to believe that every mathematical theory they come up with relates to something physically real, but the truth is that some do not relate to anything at all. One of the more famous cases of this is string theory.

The other problem with mathematicians delving into physics is that they act like a mathematical model is a complete physics theory. It is not. The science of physics requires physical explanations and mechanical models in addition to mathematical models. Look up the definition of physics if you do not believe this. There must be a physical description of how things work in any complete physics theory.

There are physical mechanisms behind every interaction. Everything moves because it is somehow pushed. In many cases those physical mechanisms are missing from popular theory. Gravity is once again a great example of this problem, as the commonly accepted theories do not include a physical model for how bodies are pushed. Acceleration requires there be a force, and a force means there is something somehow pushing on a body. Figuring out the something and somehow is critical to science, and cannot simply be ignored.

Some problems with the standard model begin with questions that are not even asked anymore. That is the case with the most fundamental questions in mechanics, what is inertia and how does it work? Another is, what causes a photon to have a rotating electric and magnetic field? And, if you do ask one of these off-limits basic questions, you risk being branded a crank, particularly if you dare to make any attempt to actually answer the question. Any theory of mechanics is incomplete without knowledge of the mechanics of inertia. Any theory of the photon is incomplete without describing the cause of the rotating electric and magnetic fields.

Another famously unasked question is, what is the force that prevents an electron from falling into a proton? If the only force at play were electrostatics, the positive and negative charges of the proton and electron would cause them to rapidly come together. If that was what happened, hydrogen would not exist, nor would any other atom or molecule. All free protons and electrons would come together to form neutrons. Every star would be a neutron star.

There is also a broad class of problems related to determining which things are elementary and which are not. Guess what, if something can be modeled out of things that are more elementary, than that something is not elementary. A big subset of these types of questions is in particle theory. Which particles are elementary particles and which are not? Which 'particles' are actually resonant states of combinations of other more elementary particles?

Unfortunately, Nobel Prizes are not given out to people who discover a new resonant state, so physicists insist that each new resonant state they discover is a particle, in order to claim their prize. The consequence of this activity is that particle theory keeps becoming increasingly complex rather than increasingly simple, as it should. To make things worse, physicists like to invent new particles in an attempt to explain things they do not understand rather than explain them using things that are known to physically exist.

The collection of 'elementary or not' problems are a subset of a broader set of problems related to determining the correct interpretation of an experimental result when there is more than one possible explanation. A typical example of this is a particle collision experiment where one type of particle is fired at another type of particle and the experiment shows no evidence of scattering. Does that mean that the target particle is tiny compared to the energetic particle, or does it mean that the target particle is larger than the accelerated particle and is transparent to it? If there is other experimental evidence that the target particle is truly large, do you simply ignore evidence, so you can continue to believe it is tiny?

This is an example of the lapses in critical thinking that are common among physicists. In many cases there is more than one explanation for an experimental result and physicists continue to insist that the wrong interpretation is the right one, even in the face of scientific evidence to the contrary.

The next class of problems occurs when physicists forget to do their job and simply call a property intrinsic. An intrinsic property is a property of a thing independent of other things. That sounds pretty innocent doesn't it? For physicists, however, it is an abdication of their responsibilities to explain in physical terms what is going on.

Often these intrinsic properties relate to the most fundamental principles in physics. Some examples are; inertia is intrinsic to a body of matter, the speed of light is intrinsic to a photon, and mass is intrinsic to particles. In order for the science of physics to progress, these types of properties must be physically explained. In other words, physicists need to do their jobs.

An even wilder set of problems occurs when there is no physical evidence at all. Modern physics is failing to draw a line between physics and science fiction. Science fiction and fantasy are fun to think about, and can make for interesting books, movies, and television shows, but they do not belong in a serious discussion of physics. The current science fiction branches of physics include many things for which there is no physical evidence, and in some cases, there is no possibility that there ever could be any physical evidence. Some of the more common examples of science fiction in physics are the many worlds and multiverse theories.

All these problems have collectively put physics in a very sad state indeed. Many important and well-regarded theories are obviously incorrect but physicists do nothing about it. Some physicists recognize the problems, but do not talk about them, while others buy into the cover-up. Most are justifiably worried for their jobs, or having to answer to the public and the people who provide funding. Many are worried that their life's work will amount to nothing. So, they keep on pretending that everything is fine with their standard model of the universe in the hopes that the problems will miraculously work themselves out.

Physicists have failed to recognize an obvious truism that relates to their situation. In order to fix a problem, you first have to admit there is a problem. That is where this book comes in. Most physicists within academic communities are unwilling to admit the problems, so they obviously need help from outsiders, physicists and others who are not entrenched in the mainstream standard model.

Physicists are obviously unwilling to question the work to which they have dedicated their lives, even if they know in the recesses of their minds that their theory is far too complicated to survive the grand simplification of physics that is coming. Nobody wants to be relegated to the dustbin of history. And yet, that is what is going to happen to most of them if they continue to do nothing about it.

I have heard it said that physicists would only address the problems with their theories when the entire world is laughing at them.[2] So please join thousands of dissident scientists in laughing at them as you read about the 100 greatest lies in physics. Perhaps then we can get on with the business of simplifying physics.

Note: The lies are not ordered by severity, but rather presented in an order which hopefully allows us to see how lies are stacked on top of more lies. As it is with any attempt at lying, lies beget more lies. Lies that did not make the top 100 are shown in bold along the way and indexed in the back of the book. My previous book The Zero-Point Universe is a good reference as it goes into some of these problems and my proposed solutions based on quantum field theory in greater depth. So, do not hesitate to go to my previous book or the Internet to look up something.

This is the 2nd edition which includes the results of research conducted for my latest book *Goodbye Quarks: The Onium Theory*.

[1] P.A.M. Dirac, "The Evolution of a Physicist's Picture of Nature," Scientific American 208:5 pp 45-53 (May 1963)
[2] F. Bishop, "The Science of Censorship," 19th Natural Philosophy Alliance Conference, Albuquerque, NM, 2012.

Lie #1: There is no Aether

> *The only equation by which the observed phenomena are satisfactorily accounted for is that of Planck, and it seems necessary to imagine that, for short waves, the connecting link between matter and ether is formed, not by free electrons, but by a different kind of particles, like Planck's resonators, to which, for some reason, the theorem of equipartition does not apply. Probably these particles must be such that their vibrations and the effects produced by them cannot be appropriately described by means of the ordinary equations of the theory of electrons; some new assumption, like Planck's hypothesis of finite elements of energy will have to be made.*[2]
> Hendrick Lorentz, 1906

Of all the lies told by physicists perhaps the greatest is that there is no aether. Since the beginning of scientific endeavor, scientists had always assumed that there was aether in space, a medium for the transmission of light and forces. Rene Descartes thought gravitational forces were transmitted by aether vortices that filled space in the 1600s and aether theory has remained with us to the present day.

While unifying electromagnetic theory, James Clerk Maxwell visualized a vortex filled aether that was responsible for the transmission of those forces and the transmission of light.[3] His visualization of rotating aether interacting with moving electric charges and producing magnetic fields was critical to his understanding of the electromagnetic equations and how they related to each other. Sadly, this history lesson on vortex theory, though not technically correct, is seldom taught in school or mentioned in textbooks.

Equally unfortunate were ill-conceived experiments such as the one by Michelson and Morley that cast doubt on the aether models of the mid to late 1800s.[4] Scientists of the day had become proficient in understanding the laws of ideal gases, including how gases interact with bodies of matter, transferring heat and energy. They naively thought that aether, whatever it may be, would similarly behave like an ideal gas.

This brought up the question of how did stars and planets not vaporize from all the heat energy transferred from the aether? It was thought that aether could produce drag on light, much like air on sound. That brought up the idea that we should be able to measure a difference in the speed of light by measuring the speed in two different directions relative to the aether.

The doubt over the existence of aether took hold at the very time when Max Planck came up with a theory of the quantum harmonic oscillator in a series of papers between 1900 and 1911.[5,6,7] His theory explained that in a system such as an ideal gas, the gas molecules never reach a state of zero energy. They always have a small amount of residual energy even when all the heat is removed from a system.

Hendrick Lorentz was perhaps the first to propose in writing that otherwise empty space may be filled with Plank oscillators and that these oscillators could be responsible for the action of all forces through the vacuum of space.[2] There is little doubt that this was a topic of conjecture between he and his colleagues.

Lorentz also recognized that these oscillators would not follow the theorem of equipartition, which in part means that the motion of these oscillators does not contribute to the heat of the system. There is no heating due to Planck oscillators in a vacuum, as the oscillators do not transfer heat energy to other bodies of matter. The decades old problem of how planets and stars are not vaporized by the aether was solved. It turns out that this also solves the Michelson-Morley problem.

Thus began the modern era of aether, often referred to as quantum field theory, quantum fluctuations, or zero-point energy. Even Albert Einstein jumped in before he became an unrepentant aether denier, by co-inventing the name zero-point energy (*nullpunktsenergie*) in a paper he co-authored with Otto Stern.[8] To be fair, they were discussing the zero-point energy of a gas rather than quantum fluctuations.

What should have been a turning point for physics, an era of real understanding of the nature of the vacuum, became a nightmare when quantum fluctuation deniers squashed those ideas and ultimately dominated physics for more than a century. Without the physical medium of the quantum field, they had to fabricate all manner of fanciful and ultimately incorrect theories to explain how light and forces are transmitted through space.

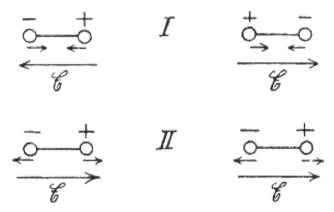

Figure 1-1 Peter Debye's illustration of pairs of dipoles shown in opposing charge orientations and showing how dipoles affect nearby dipoles depending on their charge orientation.

Then midway through the last century Hendrik Casimir and Dirk Polder produced a theory that predicted that quantum fluctuations could be detected.[9] They recognized that quantum fluctuations interact in a manner consistent with van der Waals forces.

Van der Waals forces are a collection of forces that occur between electrically charged dipoles. Each dipole has a positive side and a negative side and they oscillate in response to each other. Peter Debye was responsible for discovering one type of van der Waals force that he illustrated as shown in figure 1-1.[10]

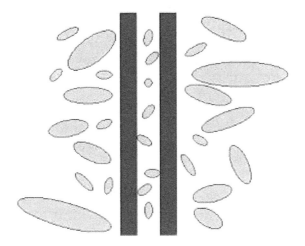

Figure 1-2 A simple illustration of the Casimir Effect between two plates. The force between two plates is reduced because longer wavelength Planck oscillators are excluded from the cavity between the plates.

Casimir's recognition that quantum fluctuations behave as electric dipoles was brilliant, and he deserved to have won a Nobel Prize for it, but sadly did not. His theory is simply explained by the two-plate illustration in figure 1-2. Quantum fluctuations exist throughout space and come in all sizes based on their wavelengths. Certain larger wavelengths do not fit between the two plates. This creates a differential in the pressure on the plates due to van der Waals forces. That is another big concept promoted by Casimir, that the quantum fluctuations exert pressure on bodies of matter.

Casimir predicted that this pressure differential pushes the two plates together, which may be the single biggest contribution he made to science. It took nearly 50 years, but experiments finally proved that Casimir was correct.[11,12] So not only do quantum fluctuations exert pressure on a body, they also cause them to move. And, they do not heat up those bodies in the process. **The**

idea that aether heats bodies is a lie. The idea that aether causes kinetic energy losses is a lie.

Decades of calculation and experimentation have shown the Casimir effect is real and additionally that Casimir forces are identical to van der Waals forces.[13] More importantly the Casimir effect proves that quantum fluctuations are real. Aether, of the type described by Planck's theory, is real.

Two other concepts that are important are: aether is composed of electric dipoles, and those dipoles interact in accordance with van der Waals forces. Dipoles also rotate, mimicking the vortex theories of old. Descartes, Maxwell, and all the others who envisioned the aether of space as vortices, were not that far off. It turns out that aether was something even simpler, rotating dipoles.

Unfortunately, even today while many physicists are now realizing that the denial of the existence of aether was indeed a lie, they have not yet come to terms with all the damage it caused, nor made any significant attempts to fix the damage.

[2] H. A. Lorentz, *The Theory of Electrons*, 2nd Edition, B. G. Teubner, Leipzig, G. E. Stechert & Co., New York, 1916, p45. (Taken from a series of lectures presented in 1906.)
[3] J.C. Maxwell, "On Physical Lines of Force" Part III "The Theory of Molecular Vortices Applied to Statical Electricity" Philosophical Magazine XXIII, 1862.
[4] A.A. Michelson, E. W. Morley "On the Relative Motion of the Earth and the Luminiferous Ether". American Journal of Science **34**: 333–345 1887.
[5] M. Planck, "Zur Theorie des Gesetzes der Energieverteilung im Normalspektrum," Verhandl. Deutsch. Phys. Ges, p.237 (1900)
[6] M. Planck, "Über das Gesetz der Energieverteilung in Normalspektrum." Ann. Physik 4, 553, 1901.
[7] M. Planck, "Eine neue Strahlungshypothese," Verhandl. Deutsch. Phys. Ges. 13: 138 (1911)
[8] A. Einstein, O. Stern, "Einige Argumente fuer die Annahme einer Molekularen Agitation beim absoluten Nullpunkt." Ann. Phys. 40: 551. (1913).
[9] H. B. G. Casimir, and D. Polder, "The Influence of Retardation on the London-van der Waals Forces", Phys. Rev. 73, 360-372 (1948).
[10] P. Debye, "Die van der Waalsschen Kohäsionskräfte." Physikalische Zeitschrift, Vol. 21, pages 178-187, 1920.
[11] S. K. Lamoreaux, (1997). "Demonstration of the Casimir Force in the 0.6 to 6 μm Range". Physical Review Letters 78: 5. doi:10.1103/PhysRevLett.78.5.

[12] U. Mohideen, A. Roy, (1998). "Precision Measurement of the Casimir Force from 0.1 to 0.9 μm." Physical Review Letters 81 (21): 4549. doi:10.1103/PhysRevLett.81.4549.

[13] K. A. Milton, The Casimir Effect: Physical Manifestations of Zero-Point Energy," World Scientific, 2001, pg. 79.

Lie #2: Michelson & Morley Disproved Aether

The second of the greatest lies in physics is that Michelson and Morley disproved the existence of aether. Anyone who has ever attended a basic physics class, or watched a science program that mentioned aether, has probably heard this lie. Of course, after reading that 'there is no aether' is a lie, it is obvious that the Michelson-Morley experiment failed to disprove it.

There are important questions related to the experiment's failure, including:

    a. Why did the experiment fail?
    b. What was wrong about the assumptions that led to the experiment?
    c. What did the experiment actually prove?

To figure out what went wrong we have to look at Maxwell's conjecture, as it was his ideas about aether and light that led to the experiment.[14] His conjecture was that there is an aether drag on light and since the earth is moving relative to the aether, he thought this drag could be detected as a change in the speed of light.

The experiment was then designed to measure the speed of light in two directions at 90-degree angles to each other. One funny oversight of the experiment is that they were thinking they could see the velocity of the earth relative to the sun, rather than perhaps the sun relative to the galaxy, or better yet the galaxy relative to the aether rest frame. The velocity of the Earth relative to the aether rest frame is what they would really measure if aether drag were real and they could actually measure it.

To digress a moment, some of you may be thinking that there is no aether rest frame. If we think about it critically, however, if there is aether it must have a rest frame, more on that later. And by the way, we have detected it and measured our velocity relative to it, as the aether rest frame is the same as the cosmic microwave background rest frame. Our velocity has been measured as 369 kilometers per second.[15] **The statement that there is no aether rest frame is a lie.**

Continuing, the idea of aether drag came about because Maxwell and others thought that light propagates through the aether in a similar way to sound traveling through air. Think about how the noise from a car has a higher pitch coming toward you than as it drives away. This is due to the Doppler effect. The speed of sound is much faster than most wind on earth, but a steady tailwind allows sound to propagate a little faster and a steady headwind makes it propagate a little slower. This is because the speed of sound relative to air's rest frame is more or less constant. A cross wind also causes sound to slow versus the propagation rate in still air, as illustrated in the following figure.

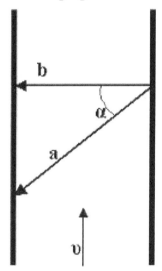

Figure 2-1 A beam of light crosses between two walls along path **b**, but because of the aether direction and velocity **v** shown by arrow, the effective distance is thought to be **a**. The angle **α** is shown between the two paths.

It is also important to note that the speed of light in Maxwell's theory achieves its maximum when measured in the aether rest frame.[14] Maxwell is credited as the first to recognize this fact. If an observer, such as on Earth, is moving relative to the aether rest frame, the speed of light in that frame of reference appears to be slower when viewed by an observer in the aether rest frame. There will be more on that later. The science of performing transformations of measurements between moving rest frames was given the name relativity. Physicists did not have a good grasp of relativity theory at the time of the experiment which led to many misconceptions.

The Michelson-Morley experiment showed that the classical concept of aether drag is a lie. The experiment did not fail; it was the initial assumptions that were wrong. Physicists misinterpreted the result and decided there was no aether, when the correct way to interpret the result was to say that Maxwell's conjecture was wrong.

Did you ever wonder why the premise of the Michelson-Morley experiment was to test for the existence of aether, but the answer was the frame transformation equations of relativity theory? One of the important things to note with the Michelson-Morley experiment is that the light source and detectors are in the same reference frame, so under any hypothetical frame transformation scheme, a frame transformation is not required for the apparatus itself.

Even more interesting, we do see energy shifts, red-shift or blue-shift of light depending on whether the emitter and detector are moving apart, or moving toward each other. But the emitted energy is independent of direction relative to the aether. So instead of seeing velocity shifts, as Maxwell predicted, we see energy shifts. And instead of seeing a directional dependence, there is none. The permittivity and permeability of space do not have a directional dependence, and since the speed of light is derived from the permittivity and permeability, it does not have a directional dependence either.

That said, light photons do have their own rest frame, the aether rest frame. You can imagine yourself riding on a photon at the speed of light. Under Maxwell's theory you would be traveling at the speed of light versus the aether rest frame. We can think of the aether rest frame as the natural photon reference frame, and the photon's energy in the aether rest frame is its natural energy.

Another major error in Maxwell's conjecture is that he thought that a photon's velocity changes are dependent on the velocity of the source relative to the aether rest frame. The detector's velocity does not matter. **It is a lie to say that light travels in the frame of reference of the light source.**

The aether, when viewed from the aether rest frame, is uniform in all directions along with its permittivity and permeability, so the velocity of photons is uniform in all directions. There is no directional dependence of light velocity as photons are always traveling in the aether rest frame, which has no directional dependence. Permittivity and permeability are uniform in the aether rest frame. **The concept of a directional dependence of the speed of light is a lie.**

When we do see aether drag from the perspective of a moving detector, we do not see it in terms of a velocity shift; we see it as an energy shift. The Michelson-Morley experiment does not show an energy shift because the source and detector are in the same frame of reference.

So, what did the experiment actually prove? It proved that:
1. Physicists did not recognize that light always travels in the aether rest frame regardless of the velocity of its source.

2. Physicists did not recognize that permittivity, permeability, and light's velocity are independent of direction.
3. Physicists did not know how to correctly perform and interpret relativistic frame transformations.
4. Physicists did not know that the observer's velocity relative to the aether rest frame causes energy changes responsible for red-shifts or blue-shifts, rather than velocity changes.
5. Physicists did not recognize that Maxwell's conception of aether drag was wrong.
6. Physicists did not recognize that Maxwell's conjecture was wrong.

With light always traveling in the aether rest frame, relativistic transformations are solely required to explain the effects of a moving source or detector. Sadly, some physicists continue to test Maxwell's conjecture and repeat the Michelson-Morley experiment even though his conjecture was wrong and the idea that light travels in the frame of reference of its source was wrong.

The statement that Michelson and Morley disproved aether is a lie. So what is really going on with the velocity of light? Well that gets us into a whole bunch of other lies.

[14] J. C. Maxwell, "On a Possible Mode of Detecting a Motion of the Solar System Through the Luminiferous Ether." Nature, 1880, Vol. XXI, pp. 314, 315.

[15] C. Lineweaver, et al. "The dipole observed in the COBE DMR four-year data.'" Astrophysics J. 470, 38,1996.

Lie #3: Wave-Particle Duality

An ocean traveler has even more vividly the impression that the ocean is made of waves than that it is made of water.[16]
Sir Arthur Eddington, 1927

The third of the greatest lies in physics is wave-particle duality. Indoctrination in the concept of wave-particle duality starts in the earliest physics classes and spreads by popular science shows, magazines, and websites. Teachers spread the dogma that a particle is not just a particle, but is also a wave. They say it is sometimes easier to treat it as a particle and sometimes it works out better to treat it as a wave. Once again, the quantum fluctuation deniers have led us astray.

How did this happen? To start, we can think about waves in water or sound waves in air. A boat moving through water creates a wave. But the boat is not a wave; the boat is just a boat. The wave is made of water. First the boat pushes the water and then that water pushes more water. The actual waves are entirely composed of water. There is no wave-boat duality.

Sound waves are similar. Imagine a bell is rung and it vibrates and sends sounds through the air. The bell is not a wave; the bell is just a bell. As it vibrates it pushes against the adjacent air, and then those air molecules push against other air molecules. The sound waves are a property of the air. There is no wave-bell duality.

So how did physicists get things so screwed up? Imagine if you would, a strange planet much like our own but the inhabitants could not see or feel the water in the oceans. They could see fish, boats, and various sorts of flotsam strangely suspended, but not the water. And then, a bright minded young physicist noticed patterns in the sand beneath the boats and fish and recognized that these patterns have certain properties as they undulate and show interference patterns. Oh and by the way, when someone suggested there was a substance they called 'water' as a possible reason for these interference patterns, the leadership on this world decreed that there is no such thing as water, as it is invisible and a famous scientist once conducted an experiment to disprove its existence.

So where do these undulating ripples come from asked our young physicist? I know, he thought, the fish and the boats and the flotsam are not just bodies of matter, but they have waves attached to them as an integral part of their make up, so sometimes you have to treat a boat or fish as a wave in order to explain the ripples in the sand. He thought these waves extended outward from each boat and fish. These wave properties interact with each other to form complex patterns in the sand, yet he still could not see the waves.

The physicists of planet Earth are sort of like that. They look at particles and wave phenomena produced by their motion, and they wonder where the wave properties came from. Being good quantum fluctuation and aether deniers, as they have been taught, they are stuck. With no medium to transmit the waves they had to come up with the crazy idea that the particle is both a particle and a wave at the same time.

This was the birth of the wave-particle duality lie. First physicists rejected the existence of aether, which would be like a boat captain denying the existence of water, or a musician denying the existence of air, but they still had to ask; where do the waves come from? They were convinced of the lie that space was empty, and thus were stuck with trying to explain how a wave could travel through empty space. So, they decided that the easiest way out of this problem was to con everyone into believing that waves are a property of particles, and that somehow particles extend infinitely through space producing wave phenomena. And at other times, the particle is just a particle and acts like it is very small.

Of course, the truth is far simpler. Quantum fluctuations do exist. The aether filling all space is real. This aether, as discussed in chapter 1, behaves like electric dipoles that move in response to moving charges. They also move in response to moving bodies of matter, which will be discussed in more detail later.

As particles move through space or rotate, they cause nearby quantum fluctuations to move and rotate. Those in turn cause other quantum fluctuations to move and rotate. Very quickly, wave phenomena are seen in the sea of quantum fluctuations as they interact with other particles or bodies of matter.

The situation is somewhat more complicated than sound waves in air or waves in water, as quantum fluctuations are not seen directly and dipole interactions are somewhat more complex. It is like the fictional planet with the invisible water that does not feel like anything. We only see the evidence that is left behind.

Wave-particle duality is a lie. A particle is just a particle. The waves are made of aether, and the aether is made of particles, so in the end, even the waves are made of particles.

[16] A. Eddington, Gifford Lecture at the University of Edinburgh, March 1927. in The Nature of the Physical World, 1929, reprint 2005, pg 242.

Lie #4: Photons are Elementary

The fourth of the greatest lies in physics is photons are elementary particles. For more than 100 years, photons have been considered elementary particles so this lie will surprise most people. Every textbook and every table of particles says it and it is repeated in every physics class, popular shows, and literature. But what they do not tell you, unless you dig deeper into the science, is that there is a more fundamental model of the photon that has been around since at least the 1930s. Those who study basic quantum field theory, Louis De Broglie, or Feynman diagrams are likely to come across it, but most fail to consider its importance.

There is a contradiction in physics teaching that begins with the most basic description of a photon. Photons are consistently referred to as elementary particles while at the same time it is said they are known to produce electric and magnetic fields, are polarizable, and have wave properties. The electric and magnetic fields, polarizability, and wave properties are true as they are consistent with experimental observation.

The first thing we notice is the wave problem. As one would expect after the last chapter, physicists have lied about photons having wave-particle duality. They say that the particle part of the photon magically produces a wave that stretches out to infinity. This wave produces the interference patterns and other phenomena that are so familiar to us. But, since we know that aether exists, the waves must be made of aether and the particle is just some type of particle, or perhaps something that just looks like a particle.

As for the other part of the description, we have the question, how does a photon produce electric and magnetic fields? Physicists do not tell us, as of course they do not know, as they failed to recognize its importance when it came up in class, or they read it in a book.

Even though photons are known to produce electric and magnetic fields they are generally thought of as being electrically neutral, and it is true that photons have zero net charge. But how does a particle with zero charge produce a rotating electric and magnetic field? How is it polarizable? It must be physics magic.

Even without the knowledge of how a photon produces electric and magnetic fields and waves while being polarizable, physicists insist that photons are elementary. Perhaps some of them do question the idea of magical propagation of fields and waves from time to time, but that sort of questioning is rapidly discouraged, as it is detrimental to all sorts of physics theories, as we will see. If you ask the question in public, or worse, attempt to answer it, you hear a chorus of 'crank' ringing out from the physics community.

If we ignore all the dogma and scientific politics, we must face the question; how does a photon produce a rotating electric and magnetic field? **The idea that electric and magnetic fields are intrinsic to photons is an obvious lie.** It turns out that the simplest way to produce an electric field is with an electric charge dipole. An electric dipole has a positive charge on one end and a negative charge on the other. If these charges are opposite but equal, the dipole still appears to be electrically neutral from a distance.

The electric field of the photon varies with time, slowly getting stronger in one direction until it peaks and gets weaker, and then it becomes negative, peaks, and returns to zero. It goes through this cycle once each wavelength. This effect can be reproduced if you have rotating dipoles where the first rotates one way, and second rotates the other way. In this manner a series of rotating dipoles produces the rotating electric field we see from a photon.

If a dipole rotates it also produces a magnetic field, so a rotating dipole simultaneously produces an oscillating magnetic field to go with the oscillating electric field. At the same time, a photon that contains a dipole is polarizable and a dipole model provides an explanation for the polarization of light.

Experimental evidence confirms that during each successive half-wavelength of the photon, the electric and magnetic fields are oriented in opposite directions. Consequently, each successive dipole must be oriented in the opposite direction from the previous dipole. This motion negates the angular momentum of the previous dipole giving a photon net zero angular momentum.

The simplest way to physically model the electric and magnetic fields of a photon is if it is composed of a series of rotating dipoles, with a new dipole appearing each half wavelength as shown in figure 4-1. This, in addition to the aether wave, is the basis for a true fundamental description of a photon as a composite particle.

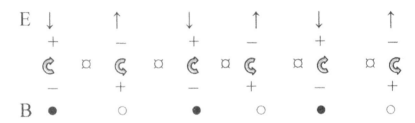

Figure 4-1 A photon as a series of rotating dipoles producing rotating electric and magnetic fields.

The idea that the electric and magnetic fields are the result of a rotating dipole is quite obvious and not new. In the early 1930's shortly after Anderson's discovery of the positron it became apparent that a photon could be described as being composed of a series of quantum electron-positron pairs. There is one quantum electron-positron pair each half wavelength. Louis De Broglie was the first to publish an electron-positron model of photons.[17]

In this case the term 'quantum' means that the electron-positron pair is a quantum fluctuation rather than a longer-lived particle pair. This is important, as a photon has no mass, and a longer-lived particle pair has mass, while a quantum fluctuation does not. In principle, a photon could be made of any physically real quantum particle pair that forms an electric dipole. Sometimes the word 'virtual' is used to describe quantum fluctuations but the word 'virtual' in that context in no way implies that the particle pair is not real.

The electron-positron model of the photon was popularized more recently by Richard Feynman and is sometimes seen in Feynman diagrams.[18] In figure 4-2 a photon is traveling from left to right represented by a wavy line, and an electron-positron pair is represented by the oval. The electron is moving in the same direction as the photon, while the positron is moving in the opposite direction.

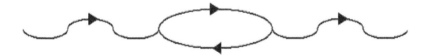

Figure 4-2 A quantum electron-positron pair as part of a photon (not to scale).

To assure ourselves that a quantum fluctuation model of a photon is appropriate, we also have to consider the energy of each half wavelength of a photon. This energy happens to be Planck's constant times the frequency divided by two, which also happens to be Planck's energy for a quantum harmonic oscillator, a quantum fluctuation. It also happens to be the maximum energy under Heisenberg's uncertainty principle, which allows a dipole to exist, but not have all its properties simultaneously detectable. Note that the field carries half the total energy separately from the dipole, but this field energy is often ignored.

The dipoles that make up the photon, and are responsible for producing the rotating electric and magnetic fields, are quantum fluctuations. Each successive quantum fluctuation appears out of and disappears into the vacuum. Each quantum fluctuation dipole is induced from the vacuum, conserving energy and linear momentum, and each successive dipole counter-rotates so that angular momentum is conserved.

The question of the composition of the electric and magnetic waves that travel with a photon is easily answered. Those waves travel through and are made of the aether, the quantum fluctuations. Each quantum dipole that is part of the photon causes adjacent quantum fluctuations to rotate. Those quantum fluctuations influence the ones next to them. Those wavelike interactions then propagate out toward infinity.

It is very clear then that photons are not fundamental as there is a more fundamental description for them as first described by De Broglie. Photons are a composite particle composed entirely of aether. Photons are composed of quantum fluctuations and are nothing more than a means of transporting energy through space. **Not only are they not fundamental, the lie is even bigger, as they are not even particles.**

To be complete, we can consider that photons could be made of other types of quantum fluctuations, and not just quantum electron-positron pairs. Any real quantum particle pair dipole with zero total charge will do. So more generally we can say, photons are composed of any real quantum particle pairs. The term 'real' here means true elementary particles, rather than composite or fictional particles.

It is also important to note again that quantum particle pairs do not exist long enough to have rest mass, so that is not an issue. The composite model of a photon does not have rest mass, just like the photon. Mass will be discussed in more detail later.

How did this lie happen? Why has this lie been perpetuated while a more fundamental model existed? Those are tough questions for physicists to answer. Ultimately, as physicists, we do not have a choice in the matter. The rotating electric and magnetic fields are physically real, and our science demands that we come up with a physical explanation for them. We cannot simply ignore them and say we do not like the way the theory or the math works. We have to make the theory work.

The statement that photons are fundamental is a lie. They must be composite particles in order to explain the rotating electric and magnetic fields.

[17] L. De Broglie, Matter and Light the New Physics, translated by W.H. Johnston Dover Publications, pg. 100101, 1939 (French edition Matiere et Lumiere, 1937).
[18] P.W. Milonni, The Quantum Vacuum, Academic Press LTD, London 1994, p. 49.

Lie #5: Virtual Photons

The fifth of the greatest lies in physics is the virtual photon. A virtual photon is a type of quantum fluctuation made from light photons. This is one of the more significant lies in that it has slowed progress toward understanding aether theory. Based on the last chapter, some reasons why virtual photons are not real should already be apparent.

In most theories about quantum fluctuations, they are modeled as particle pairs, which can be produced out of the vacuum, and annihilate and returned to the vacuum These particle pairs meet the Planck energy condition and do not exceed the Heisenberg limit. Each quantum particle pair consists of a matter particle, along with its opposite antimatter particle. Photons are said to be their own anti-particle, so a virtual photon is two photons. This highlights another lie. **If photons are not a particle, photons cannot be an antiparticle either.**

When it comes to modeling the energy of the quantum field, physicists usually model it using the most basic particle they know, the photon. Thus, virtual photons are frequently stated to be the basic form of zero-point energy. This incorrect supposition is a big problem for physics.

While photons do have an electric and magnetic field, those fields appear electrically neutral when viewed from a distance. Virtual photons are then seen as electrically neutral. That said, some physicists are smart enough to recognize that due to the electric and magnetic fields, photons can still be treated like dipoles, and they still interact in ways consistent with van der Waals forces. But, in the vast majority of theories, the dipole character of virtual photons is ignored.

To understand the true depth of the problem we have to look at the energy of a virtual photon pair. If we consider the energy of two photons appearing from the vacuum and existing for one full wavelength out and another full wavelength back, that yields a combined energy of **4hv**, where **h** is Planck's constant and **v** (the Greek letter nu) is the frequency. Virtual particles, however, cannot exceed **½hv** since that is the energy of the Planck oscillator, and the time-energy limit according to Heisenberg, keeping in mind that there is another **½hv** in the field. Therefore, a virtual photon cannot be composed of two full wavelength photons.

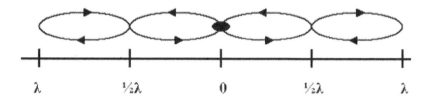

Figure 5-1 A quantum fluctuation as a photon pair originating at the center with each photon traveling out one full wavelength and then back one full wavelength. The arrows show the direction of propagation. λ (lambda) is the symbol for wavelength.

Next we can consider each half of the photon pair going out a half wavelength and returning. In that case the virtual photon has **2hv** energy, which is still too much. Instead of going out and back we could think of a virtual photon as two half-wavelength photons side by side, but somehow not returning. In that case the energy is still **hv** and not **½hv**.

If instead we try to go to two quarter-wavelength photons, they cannot exist. A quarter-wavelength reflector prevents that wavelength from existing in that cavity, as it interferes with itself. Quarter-wave coatings for example, are used in lenses to prevent reflection.

In order to have a virtual photon that has the Planck energy, it must be a single half-wavelength photon and not a photon pair at all. In that case, however, you cannot have photon pair production, or photon pair annihilation, as there is no photon pair. If there is no opposing pair, it is not a quantum fluctuation, as it does not start and end with zero energy. The single photon model for a virtual photon fails the energy conservation test.

Fortunately, there is a simple way out of this conundrum. As discussed in the last chapter, a half-wavelength photon is more fundamentally described as a particle pair dipole, such as a quantum electron-positron pair. A quantum electron-positron pair allows for pair production and annihilation and matches the energy limits for a quantum fluctuation. It should be no surprise that the electron-positron model for a photon is also the proper fundamental description of a virtual photon, which is not really a photon at all.

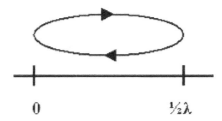

Figure 5-2 A quantum fluctuation as a single photon of a half-wavelength. The arrows show the direction of propagation.

It is bizarre that physicists speak about virtual photons, but fail to perform a basic check to see if they can actually exist. It is simple to see that every viable photon pair model has too much energy to be a real quantum fluctuation.

Just like with photons, virtual photons, to the extent they exist, can be any real quantum particle pair. And yes, they are massless, as they do not exceed the Planck energy.

The concept of the virtual photon is a lie. Aether physics would advance far more rapidly if physicists would only recognize this simple fact.

Lie #6: Aether is Virtual Photons

The 6th great physics lie is aether is virtual photons. This lie is, of course, a direct consequence of lies #4 and #5. All forces must somehow be transmitted through the aether, so understanding the composition of aether is critical to understanding how forces work. Because physicists described aether incorrectly, if they were not busy ignoring its existence altogether, they have stalled the progress toward a better understanding of force theory

Physicists who were not complete aether deniers had to determine the best way to model quantum fluctuations. In order to not violate the principle of conservation of energy, quantum fluctuations have to start and end with zero energy and cannot exceed Planck's energy limit. So far, the only way we have to model quantum fluctuations such that there is zero energy before and after is to assume they are matter-antimatter particle pairs. These pairs are produced together and annihilate each other. That part of the model is easy.

The next step was deciding what type or types of particle pairs would make up the aether. Mainstream physicists improperly chose the photon since it was thought to be the most basic aether particle. The photon was also falsely thought to be its own antiparticle, when neither is really a particle. So, physicists screwed up and decided to model the quantum field as virtual photon-antiphoton particle pairs.

Since photons are thought of as electrically neutral when viewed on a large scale, aether was also viewed as electrically neutral. Most physicists did not think about the possibility that aether acts like dipoles even after Casimir suggested it. Oh, if they thought about it, they would realize that since photons have rotating electric and magnetic fields, they at least act like dipoles, but very little work was done exploring the consequences of the dipole character of photons.

Zero-point field theory has languished under the virtual photon model. Even though photons produce electric and magnetic fields, they were not thought to have an important contribution to electromagnetic force theories, other than as a gauge boson or force carrier, a particle that was somehow responsible for communicating forces. As for the strong, weak, and gravitational forces, virtual photons were only thought to participate in interactions in a minor way, if at all.

Now we can see that aether cannot be made of virtual photons because virtual photons do not exist. Aether must be made of other types of quantum particle pairs.

The choice we make is important. If we choose the wrong types of particles, we get additional errors. As we progress through this book, we will learn about more lies as they relate to other particles in the standard model which are not real and cannot be part of the aether. For now, we can start with the most obvious aether constituent, the quantum electron-positron pair, with the allowance that aether can be made of combinations of any real quantum particle pairs.

With the photon out of the way, the electron and its antimatter opposite the positron are the most fundamental of all particles. They are also permanently stable, which makes them very special, as only two particles are. While there is an equation that describes some of the electron's characteristics, the Dirac equation, there is currently no physical description of an electron that would make it something other than an elementary particle.

It may be that the quantum fluctuations are not really electron-positron pairs, but rather some precursor of electrons and positrons. Free and stable electrons and positrons have mass, while quantum fluctuations do not. That leads to two important questions; how does that happen, and what is mass? The mass questions will be dealt with later. Suffice it to say for now, that when we think of quantum electron-positron pairs we should think of something electron-like and positron-like, not necessarily identical to the permanently stable versions of those particles. Initially the most important properties are that they are matter-antimatter pairs that form electric charge dipoles.

For the sake of symmetry and other arguments that will be developed later, there must also be a dipole that forms a matter-antimatter particle pair where the matter particle has a positive charge and the antimatter particle has a negative charge. With respect to the properties of matter and charge one could say this type of quantum particle pair is proton-like and antiproton-like. That is not to say that they are actual protons and antiprotons. Well enough of that for now, as we will come back to this topic as we explore more great lies.

Aether composed of virtual photons is a lie. Once we recognize that aether is made of quantum particle pair dipoles we can readily see how aether interacts with bodies of matter. The Casimir force is only the tip of an iceberg of knowledge just waiting to be gained. We will also see that the existence of dipolar aether requires that many theories be changed to adapt to its presence.

Lie #7: Photons are a force carrier

The 7th great physics lie is that photons are a force carrier. Force carriers are called gauge bosons so it is more commonly said that photons are gauge bosons. Gauge bosons are particles in the standard model that are responsible for transmitting information about forces. Two particles, or bodies of matter, exchange these particles in order to communicate.

This idea stems from the fundamental problem that quantum field deniers brought upon themselves. Without aether filled with quantum fluctuations, the vacuum has no medium that could be responsible for transmission of forces. Without quantum fluctuations, the deniers were left with action at a distance by magic. So in order to look a little less like wizards, and more like scientists, they invented the idea that forces could be carried by particles called gauge bosons. They made the photon the gauge boson for electromagnetic theory, the force carrier of electric and magnetic forces.

This is, however, only a slight-of-hand trick that relocates the problems rather than solving them. Further, it adds new, even worse problems. As we will see after a review of the problems with gauge bosons, it really does make one question the critical thinking ability of physicists.

To start with, a photon's energy, wavelength, and frequency are not independent. A photon carries one bit of data, in addition to direction, which is not enough information to act as a force carrier. A virtual photon may have a wavelength and a direction, but by extension of Heisenberg's uncertainty principle it is not possible to measure both simultaneously. So we are still stuck with one bit of information per virtual photon.

Additionally, particles do not carry the supercomputer required to process the data from a nearly infinite number of gauge bosons. There is no known physical mechanism for individual bits of data to be collected and organized into a meaningful instruction.

Also, since a virtual photon is equivalent to a half-wavelength of a photon, in order for particles to exchange a virtual photon, the virtual photon's energy is solely dependent on the distance between particles. So, the only data that a virtual photon can transmit is the virtual photon's wavelength, which equates to distance.

Then, in order for a photon to be emitted and absorbed, it cannot be a virtual photon, as there is no known physical mechanism available for a particle to emit or absorb a virtual photon.

But, if a gauge boson is a non-virtual photon, the emitting particle will lose energy and the receiving particle will gain energy, which is not consistent with observation. And, once again there is no physical mechanism in the standard model for a particle to actually produce a photon.

Also important is that particles do not have a propulsion system so they cannot move appropriately once told how to move by gauge bosons.

Most obviously, photons and virtual photons are not really particles, as discussed in Lies #4 and #5. Since photons are not fundamental particles in the first place, they are ineligible to be gauge bosons. It is the actual quantum fluctuations of the aether that are a far better candidate as a force carrier.

There is one sense in which this lie was somewhat correct, for it is sometimes stated, electric and magnetic fields appear to be made of photons. The mistake was that it was not photons, but rather quantum fluctuation dipoles that form the electric and magnetic fields. For once you have quantum charge dipoles throughout space, electromagnetic field theory suddenly makes sense.

Lie #8: Electromagnetic Fields are Not Real

> *I believe every physicist feels inclined to the view that all the forces exerted by one particle on another, all molecular actions and gravity itself, are transmitted in some way by the ether, so that the tension of a stretched rope and the elasticity of an iron bar must find their explanation on what goes on in the ether between the molecules. Therefore, since we can hardly admit that one and the same medium is capable of transmitting two or more actions by wholly different mechanisms, all forces may be regarded as connected more or less intimately with those which we study in electromagnetism.*[19]
>
> Hendrick Lorentz, 1906

The 8th great physics lie is electromagnetic fields are not real. Michael Faraday discovered electric and magnetic fields in the early 1800s. His fields were a convenient way to describe how electric and magnetic interactions propagate through space. These fields became the basis for calculations and are included in the fundamental equations of electricity and magnetism.

In Faraday's day, nobody knew what these fields were, or what produced them. There was a notion that there was something in the aether. When Maxwell unified the electric and magnetic equations into a cohesive electromagnetic theory that bears his name, he visualized the fields as vortices. He thought a current running through a wire would cause a vortex to rotate, and conversely a rotating vortex in the aether could cause a current in a wire. Somehow, he thought it possible for vortices to combine into electric fields. This model worked well enough for magnetic fields, but not so well for electric fields, and while it was important to his discoveries, that part of his theory was abandoned long ago.

There was still some idea that fields somehow propagate through aether when the Michelson-Morley experiment happened and aether theory came under question. At this point aether deniers were stuck. If there were no aether then the fields were not physically real. And yet, field theory provided a good model for visualizing how forces work, and the equations were still based on field strengths.

For over a century, young physicists have been taught about fields, how to illustrate them, calculate them, and measure them. Then they are told that electromagnetic fields are not real. They are taught that there is nothing physically in space to explain them, and thus fields are merely a visual and mathematical modeling technique that we can measure.

This is a problem. If the fields are not real, there is no way for electromagnetic forces to propagate through space. And there is no way the fields could be measured. Without them there is no medium of transmission for forces between particles and bodies of matter. And additionally, there is no action mechanism to cause particles and bodies of matter to move. Physicists were left with magical action at a distance and magical motion theories that are still with us today.

There were physicists who recognized that quantum fluctuations exist, but mistakenly modeled them as photons. They then had an idea that the collection of photons had something to do with the electromagnetic field in a way that was different from the gauge boson model. Some went so far to say that this field of photons was polarizable, yet still they fell short of recognizing that the field was made of dipoles, which certainly are polarizable.

We now know, as Casimir predicted, that the aether is filled with electric dipoles. These dipoles have a positive charge on one end, and a negative charge on the other. Based on Casimir's theory we also know that these dipoles interact in a manner consistent with van der Waals forces. Aether dipoles move depending on the motion of other aether dipoles, and can push on bodies of matter, as has been proven experimentally.

> *This leads us to a picture of discrete Faraday lines of force, each associated with a charge, -e or +e. There is a direction attached to each line, so that the ends of a line that has two ends are not the same, and there is a charge of +e at one end and –e at the other.*[20]
>
> Paul Dirac, 1963

If aether dipoles move due to the motion of other dipoles than they can also move due to electric charges, whether those charges are stationary or moving. Aether dipoles are polarized by the existence of an electric charge. Quantum dipoles rotate due to the motion of an electric charge or current.

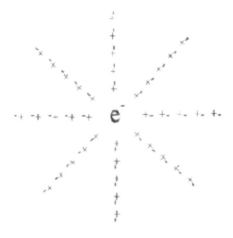

Figure 8-1 A free electron surrounded by aether dipoles that are polarized in response to its electric charge.

Electric and magnetic fields are real. They are composed of the quantum fluctuation dipoles that fill the aether. Electric fields are a polarized aether. Magnetic fields are rotations of the aether. Electromagnetic fields, whether polarizations or rotations, propagate point-to-point dipole-to-dipole throughout the aether. Electric and magnetic fields are measurable.

The teachings that electromagnetic fields are not real are lies. Electromagnetic field theory is not an abstract concept; it is physically real. Fields are not produced by magic. The dipoles of the quantum field really exist.

[19] H. A. Lorentz, *The Theory of Electrons*, 2nd Edition, B. G. Teubner, Leipzig, G. E. Stechert & Co., New York, 1916, p45. (Taken from a series of lectures presented in 1906.)
[20] P.A.M. Dirac, 'The Evolution of the Physicist's Picture of Nature." Scientific American, May 1963, 208(5), 45-53.

Lie #9: Electromagnetic Theory Explains Motion

The 9th of the greatest lies in physics is electromagnetic theory explains motion. Werner Heisenberg, along with a majority of physicists since, consider electromagnetic theory closed, a complete theory.[21] Unfortunately they never got around to describing how bodies move in response to electric and magnetic forces. They never came up with a physical model for the fields either. They neglected to ask, what does the pushing and how is that push generated? Or if they did ask, they were quickly hushed and told not to worry about such things. Present day mainstream physicists continue to act like their theory explains motion, when it does not.

How this happened is a problem consistent throughout physics. First, the mathematicians have taken over physics. Shut up and calculate is their mantra. They do not seem to even care about understanding the fundamental physical phenomena, the fundamental physics. As long as they can perform a calculation that allows them to predict some results, they are happy.

The second part of the problem lies with the quantum fluctuation deniers who control physics. When the only thing between two bodies is vacuum, and they deny the existence of anything in the vacuum, all they can do is give up on ever understanding the underlying mechanisms. Then they must lie continually to the public, and indoctrinate students in their false dogma, acting like they somehow understand. In moments of honesty, they may allow that it is beyond their understanding and being forever ignorant is OK.

Fortunately, we do not have to remain in a state of ignorance, as quantum fluctuations do exist, and they make a perfect medium for transmitting forces between bodies, and as it turns out, for producing motion.

We are also fortunate that Hendrik Casimir, one of the most brilliant physicists of the last century, figured out the fundamentals of how forces work in relation to the vacuum. He figured out that quantum fluctuations are charge dipoles that interact to produce van der Waals forces, and push bodies of matter.

He also recognized that van der Waals forces produce pressure on objects, and this pressure changes if some quantum fluctuations were excluded from a region of space. As stated previously, in his most basic example the exclusion zone is the space between two plates and these plates are pushed together when they are close enough to exclude significant quantum fluctuation wavelengths. The outside pressure pushing the plates together then exceeds the inside pressure pushing them apart.

The quantum fluctuation deniers have been too busy trying to ignore Casimir's work for the past 70 years to realize that his force mechanics can be extended to all electromagnetic forces. Van der Waals forces, and by extension the Casimir forces, are after all, a part of electromagnetic force theory.

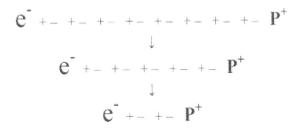

Figure 9-1 A free electron and proton being pushed together as the pressure is reduced between them.

For example, two bodies with opposite charge polarize the quantum fluctuations between them. Positive charges in the quantum dipoles point to negative charges on other dipoles as seen in the figure above. When a dipole annihilates, the dipoles next to it are brought closer together. This leads to a reduction in the van der Waals pressure between opposite charges, allowing the outside van der Waals forces to push them together.

Figure 9-2 Two free electrons with polarized dipoles between them, showing how the pressure increases in-between pushing the electrons apart.

On the other hand, when there are two like charges, the quantum dipoles at the midway point oppose each other, leading to an increase in the van der Walls pressure between the charges, thus pushing the bodies apart. The outward pressure overcomes the pressure pushing them together.

Most people have played with magnets and are familiar with how if you try to put the north poles of two magnets together, they are pushed apart. And, if you do not control them, one will rapidly spin around so that the north and south poles of the two magnets slam together.

What most people do not realize is that there are little quantum fluctuation dipoles between the magnets. These dipoles rotate so that they too become little magnets. These little quantum magnets then align with the physical magnet such that their south pole is near the physical magnet's north pole.

In this way there are a bunch of little quantum magnets lined up between the two physical magnets, and just like with charges, if all those quantum magnets are lined up north pole to south pole, the pressure between the magnets is reduced. And then, the magnets are pushed together by the outside van der Waals forces.

Then if two physical magnets are lined up with their north poles together, some of the quantum magnets have their north poles together, causing them to be diverted away from each other and thus increasing the pressure between the two magnets. The pressure pushing the magnets apart is then greater than the pressure pushing them together, and the magnets fly apart. The same thing happens when you press the south poles of two magnets together.

Mainstream physicists are certainly lying when they pretend their theory accounts for electromagnetic motion. Sadly, the simple explanation has been right in front of them for decades. Electromagnetic motion is due to the extended Casimir effect. Their denial of the existence of aether is responsible for their failure as scientists.

The statement that electromagnetic theory explains motion is a lie. Once we understand quantum fluctuations and Casimir's theory, it is obvious that all electromagnetic motion is due to pressure differentials in the zero-point field, the extended Casimir force.

[21] A. Bokulich, "Open or Closed? Dirac, Heisenberg, and the Relation between Classical and Quantum Mechanics." Studies in History and Philosophy of Modern Physics 35(3) (2004).

Lie #10: Magnetic Monopoles

The 10th great physics lie is magnetic monopoles. As most people know, all magnets have a north pole and a south pole. Even if a magnet is cut in two, both pieces still have a north pole and a south pole. The magnetic monopole idea is that somehow a north pole could exist without a south pole and a south pole could exist without a north pole.

Some physicists want to believe that magnetism can be described like electrostatic forces. Electrostatic forces have separate positive and negative charges, so it would be mathematically convenient if north and south magnetic poles could be treated like two separate magnetic charges. This is nothing more than wishful thinking with no basis in physical reality, but it continues to come up.

This is another problem based on the denial of quantum fluctuations. In particular, if one recognizes the existence of the Casimir effect, space is full of electric dipoles. These dipoles rotate to form magnetic fields, and lots of little magnets. Rotating electric dipoles always have both a north and south pole, so any magnetic field that exists, must also have a north and south pole. There are no magnetic fields independent of moving electric charges.

Even the most fundamental magnetic field, the magnetic field of an electron, has a north and south pole. The magnetic poles of particles do not, and cannot exist independently of each other.

The big bang model for the universe is also said to have a monopole problem, since some physicist's ideas of how matter formed just after the big bang, include the formation of numerous monopoles. Since magnetic monopoles do not exist experimentally, and cannot exist in a zero-point field filled with dipoles, this is a big problem. **This tells us that the physicist's formulas**

for the production of matter requiring magnetic monopoles are lies.

Magnetic monopoles are a lie. They cannot possibly exist. **All theories that include hypothetical magnetic monopoles are also lies**. Being truthful, magnetic monopoles fall squarely in the science fiction category.

Lie #11: The Speed of Light is Constant in All Reference Frames

The 11th great physics lie is that the speed of light is constant in all reference frames. We should of course add the caveat 'in a vacuum.' The constancy of the speed of light is another problem that lands at the feet of quantum fluctuation deniers. And, it ends up causing other problems, other lies, particularly with respect to understanding how different observers moving at differing velocities see light from a given source. In other words, it creates problems for relativity theory.

It was Maxwell who first had a decent grasp of a theory of light. He recognized that light was electromagnetic, and had to fit in with the collection of electromagnetic equations that he compiled. He also recognized that light traveled through the medium of the aether, the luminiferous aether as it was called. Additionally, he realized that aether had to have a rest frame, a frame of reference where aether is at rest, or more generally its motion averages to zero. In its rest frame aether is homogenous, and the speed of light, permittivity, and permeability are constant in all directions.

Most importantly, with regard to this topic, Maxwell recognized that light achieves its maximum velocity in the aether rest frame.[22] Consequently, light appears to have a slower velocity in any frame of reference moving relative to the aether rest frame. He came to this conclusion because of his hypothesis of aether drag on light, the same hypothesis that led to the Michelson-Morley experiment. While his drag hypothesis was wrong, he was not wrong about the speed of light in moving frames of reference. He was unaware, however, that clock rates also changed so the speed of light appears to be the same in all reference frames.

After the Michelson-Morley experiment, aether deniers took over, and Maxwell's theory of light velocity came under question. Under Maxwell's theory the speed of light was a property of the luminiferous aether with respect to how it interacts with light photons. If there is no aether, how is the speed of light determined? Aether deniers reached the only conclusion that they could; the speed of light had to be an intrinsic property of light. Light simply knows what velocity it is supposed to have. **Of course, the idea that the speed of light is an intrinsic property of light is a lie**. The speed of light is a function of the permittivity and permeability of the quantum field.

Since we know that aether exists and is for all intents a luminiferous aether, we have to go back to Maxwell's hypothesis that the speed of light is due to the interaction between photons and their propagation medium, the quantum fluctuations. Then we need to understand the Michelson-Morley experiment result in that context. There are four important items to consider.

1. What the Michelson-Morley experiment showed was that the two-way velocity of light in a given frame of reference is independent of direction. Note that the term two-way is used to be precise, as the experiment does not attempt to measure the one-way speed, which requires clock synchronization.
2. The speed of light is the same when measured in any reference frame regardless of its velocity relative to the aether rest frame. The speed is approximately 3×10^8 meters per second. Note here that there are two units, one for distance and one for time, as that is important.
3. A monochromatic source of light emits the same energy in a given reference frame regardless of direction. At the same time a monochromatic light will have a different energy when measured by a detector moving relative to the source. So, a detector's velocity relative to the light source affects the measured photon energy.

Based on the above evidence there are two possible conclusions regarding the speed of light.

> A. The speed of light is constant in all reference frames, and the velocity relative to the aether rest frame does not change the velocity of light.
> B. The speed of light is not constant in all reference frames, but the clock rate in a moving reference frame changes such that the velocity of light appears to be the same in a moving frame of reference.

We must consider relativistic effects on moving frames of reference to figure out which of the above two choices is correct. The clock rate in a moving frame of reference is slower than the clock rate in the aether rest frame, as has been proven experimentally. The basic equation for clock rate slowing is equation 11-1, where **t'** is the time interval in the moving frame of reference, **t₀** is the time interval in the aether rest frame, **v** is the velocity of the frame of reference relative to the aether rest frame and **c** is the speed of light.

Equation 11-1

$$t' = \frac{t_0}{\sqrt{1 - \frac{v^2}{c^2}}}$$

Equation 11-2

$$x' = \frac{x_0}{\sqrt{1 - \frac{v^2}{c^2}}}$$

The distance traveled in a frame of reference moving relative to the aether rest frame is less than in the aether rest frame per the basic equation for distance traveled in equation 11-2, where **x'** is the distance traveled in the moving frame of reference and **x₀** is the distance traveled in the aether rest frame. Note that the term in the denominator is the same in both equations. Time and distance change at the same rate.

Consequently, the speed of light is slower in a frame of reference moving relative to the aether rest frame, but it is not noticed because the clock rate is slower to the same degree. The speed of light as measured in any moving reference frame is the same as measured by someone in that reference frame, even though the speed of light is not constant when measured with a clock in the aether rest frame.

This provides another answer to the question of how the velocity of light is independent of direction. It is in part because the clock rate of the aether rest frame is independent of direction.

So Maxwell was right about the speed of light being fastest in the aether rest frame and slower in any other frame of reference, at least when measured by an observer in the aether rest frame. He was wrong in thinking that aether drag has a directional dependence, given that light always travels in the aether rest frame where the permittivity and permeability have no directional dependence. Both photon energy and velocity are independent of the direction a photon is emitted from a source.

The statement that the speed of light is constant in all reference frames is a lie. The velocity of light only appears constant because clock rates change an equal degree to make it appear that way. The speed of light in a vacuum is only constant in the aether rest frame where the photon is actually traveling regardless of the motion of the source or detector. So It really is constant, but not in all reference frames as we are told, but that is next.

[22] J. C. Maxwell, "On a Possible Mode of Detecting a Motion of the Solar System Through the Luminiferous Ether." Nature, 1880, Vol. XXI, pp. 314, 315.

Lie #12: The Speed of Light is Constant for all Observers

The 12th of the greatest lies in physics is that the speed of light is constant for all observers. Once again, we must add the caveat in a vacuum. This lie follows directly from lie #11 that the speed of light is constant in all reference frames. As with most lies in physics this one is due to the denial of the existence of quantum fluctuations, the modern aether. Without a medium of transmission for photons, the speed of light had to be a fundamental property of the photon.

When formulating his theory of special relativity, Einstein made the assumption that photons had to have the same velocity for all observers no matter what their velocity relative to each other.[23] He did not believe in aether at that time, and so did not factor the aether rest frame into his thinking; in fact, he removed the aether from other people's theories to devise his own special theory. Einstein is said to have gone through a couple of periods in his life when he thought that quantum fluctuations and the aether might be real,[24] but for the most part he rejected the existence of aether and his published theories reflect that point of view.

At this point it is important to consider the two-way speed of light versus the one-way speed of light, as this is another way some people think they can prove the existence of aether. They think the one-way speed of light would differ depending on its direction relative to the aether. Once again this is based on an aether drag hypothesis.

Since light travels in the aether rest frame rather than the reference frame of a moving source, this test is invalid. Light travels the same speed in any direction in the rest frame. Consequently, any experiment that could show the one-way speed of light is the same in two directions would not disprove the existence of aether.

The speed of light is the same in two directions since the permittivity and permeability are the same.

Scientists cannot directly measure the speed of the light coming from a star or galaxy. The one-way speed of light also cannot be measured directly without making assumptions in advance about the synchronization of clocks and relativistic clock rate changes. We cannot know experimentally whether the one-way speed of light is constant for all observers by trying to measure light speed directly.

If we consider a stream of photons from a single source, they can hypothetically be observed by a nearly infinite number of observers going a nearly infinite number of different relative velocities. All the observers will say that the photons are traveling the same speed, the usual speed of light.

This begs the question of how could a photon move a nearly infinite number of relative velocities at the same time? This question of course fascinates physicists, and they appear to have been more driven to accept it due to their fascination, rather than reject it because it sounds irrational. The more rational view would be to defer to relativity theory and say that light only appears to travel the same speed because the measurement of distance and the clock rate is different in each different moving frame of reference. The light is, of course, always traveling in the aether rest frame.

The quantum fluctuation deniers were, of course, wrong. Quantum fluctuations and aether do exist. Aether is the transmission medium for photons. Permittivity and permeability are properties of the aether, determined by the aether. Photons consist entirely of quantum fluctuations, so for photons to have infinitely variable relative velocities, all the quantum fluctuations of the aether would have to rotate in response in an infinite number of photon dipole angular velocities. Effectively we would need an infinite number of aether rest frames.

That is impossible and completely irrational. Instead, aether has a single rest frame, as has been experimentally verified by measurements of the cosmic microwave background. All light is transmitted in that rest frame.

As Maxwell concluded in the mid 1800s, the maximum speed of light is only achieved in the aether rest frame. In any other frame of reference, the speed of light is slower when measured with a clock in the aether rest frame, but it looks to be the same to the moving observer due to relativistic clock slowing of the moving observer's clock.

As noted before, since light is composed of the same quantum fluctuations that make up aether, and light travels through aether, it is proper to think of all photons traveling in the aether rest frame. That way if we actually try to calculate changes in energy, distance, or clock rate, we can use the aether rest frame as a standard frame of reference for all conversions. Even though the prevailing wrong theory has no standard frame of reference, we still have to assign one in order to perform calculations. It is better for us that there actually is a standard frame of reference. It also avoids numerous paradoxes from cropping up.

The assumption that the speed of light is the same for all observers is a lie. This lie came about due to the rejection of aether and a poor assumption of how photons travel through the luminiferous aether.

[23] A. Einstein, "Zur Elektrodynamik bewegter Körper." Annalen der Physik 322 (10): 891–921, 1905.
[24] A. Einstein, "Æther and the Theory of Relativity." Address delivered on May 5th, 1920, at the University of Leyden, Germany.

Lie #13: Special Relativity

The 13th of the greatest lies in physics is special relativity. This is not to say that all relativity theory is a lie as we have seen it is necessary for the understanding of time, distance, and energy, from the perspective of observers in moving frames of reference. The problem is that Einstein made a specific assumption when he developed special relativity which distinguished his theory from earlier relativity theories. Einstein assumed that aether does not exist. So, this is another case of an aether denier coming up with an invalid theory.

While in popular science it is commonly said that Einstein is the inventor of relativity theory, it was the French physicist Poincaré who coined the term. Many scientists made important advancements to relativity theory before Einstein, such as Fitzgerald, Lorentz, Larmor and of course Poincaré.[25-32] Those early developers of relativity theory assumed that aether exists and they took steps to incorporate aether into relativity theory.

Einstein refused to recognize that aether exists, and just as importantly, refused to recognize that there is an aether rest frame, and that the aether rest frame is the standard frame of reference for relativity theory. This led to the problem that any frame of reference could be selected as a standard frame of reference when solving a given problem. But, depending on which frame of reference is selected, there are differing results. **The idea that one can perform relativistic frame transformations without a standard frame of reference is a lie.**

These differing results led to the so-called paradoxes of special relativity, including the famous twin paradox. The paradoxical nature of special relativity should have been a warning. **Any theory that introduces unresolveble paradoxes is a lie.**

Now physicists say they resolved the twin paradox, by selecting a standard frame of reference, but it does not solve the basic underlying paradox. Under special relativity when two twins are moving apart at near the speed of light, we have no way of knowing which twin is getting older. In Lorentz relativity we measure their velocities relative to the aether rest frame, and then we know which one is getting older faster, and which one has the slower clock rate.

In light of the proven existence of quantum fluctuations, Einstein's assumptions are now confirmed to be incorrect. Here is a list of the biggest mistakes he made while developing his special relativity theory:

1. Einstein wrongly assumed that there is no aether
2. Einstein wrongly assumed that there is no standard rest frame
3. Einstein wrongly assumed that the speed of light is intrinsic to a photon
4. Einstein wrongly assumed that a standardized rest frame is unnecessary
5. Einstein wrongly assumed that the speed of light is constant in all frames of reference.
6. Einstein wrongly assumed that light travels in the frame of reference of its light source.
7. Einstein wrongly assumed that the speed of light is constant for all observers.

Einstein chose poorly. One consequence of all these false assumptions is that numerous errors crop up anytime someone tries to apply special relativity. In the end, physicists actually use the Lorentz-Poincaré version of the equations, often in an updated form, and they have to carefully select a standard frame of reference when performing transformations. They must be particularly careful when dealing with a rotating frame of reference like Earth.

Now that the existence of the Planck type aether has been experimentally proven, any physicist with a basic knowledge of relativity theory knows that Einstein was wrong and special relativity is a lie. Some physicists do say that special relativity theory is inconsistent with quantum field theory, which is just another name for aether theory, but they fail to take the important and necessary step of discrediting special relativity. Note that Lorentz-Poincaré relativity has some lies embedded in it too, as we shall see shortly.

[25] G.F. FitzGerald, "The Ether and the Earth's Atmosphere." Science 13 (328): 390, 1889.
[26] H.A. Lorentz, "The Relative Motion of the Earth and the Aether." Zittingsverlag Akad. V. Wet. 1: 74–79, 1892.
[27] J. Larmor, "On a Dynamical Theory of the Electric and Luminiferous Medium, Part 3, Relations with material media." Phil. Trans. Roy. Soc. 190: 205–300, 1897.
[28] J. Larmor, *Aether and Matter*, Cambridge University Press, 1900.
[29] W. Voigt, "On the Principle of Doppler." Nachrichten von der Königl. Gesellschaft der Wissenschaften und der Georg-Augusts-Universität zu Göttingen (2): 41–51, 1887.
[30] H. Lorentz, "Simplified Theory of Electrical and Optical Phenomena in Moving Systems.", Proceedings of the Royal Netherlands Academy of Arts and Sciences 1: 427–442, 1899.
[31] H. Lorentz, "Electromagnetic phenomena in a system moving with any velocity smaller than that of light.", Proceedings of the Royal Netherlands Academy of Arts and Sciences 6: 809–831, 1904.
[32] H. Poincaré Bull. Sci. Math, (2) 28, 317-, November 1904. English translation: Bull. Amer. Math. Soc. 37, 2000, 25 - 38.

Lie #14: Length Contraction

The 14th of the greatest lies in physics is length contraction. Length contraction implies that physical rods and other objects are shorter when they are moving relative to the aether rest frame. This phenomenon is related to relativity theory whereby an observer in one reference frame, observing objects in a second reference frame moving at a velocity that is significant with respect to the speed of light, sees a shortening of the distanced traveled. While this shortening of distance traveled does occur, the result is often misinterpreted or misrepresented as length contraction, or more problematically rod or object contraction.

This lie came about when physicists were trying to figure out why the Michelson-Morley experiment failed. That experiment failed because Maxwell and others falsely assumed that the speed of light has a directional dependence due to hypothetical aether drag. When it was shown that the speed of light, as measured in a given frame of reference, is not directionally dependant, physicists were puzzled.

Physicists should have questioned the initial assumption of directional dependence of the velocity of light but instead they went about trying to come up with an alternative way to solve the problem. George Fitzgerald proposed the solution of length contraction in 1989 and Hendrik Lorentz independently came up with the same idea a few years later, but was more successful at popularizing it.[25,26] Now it is known as Lorentz-Fitzgerald length contraction.

Their solution was that distances became physically shorter depending on the direction relative to the aether rest frame. In the context of the Michelson-Morley experiment, even though the path lengths had been measured to be the same, the length contraction theory said that they were actually different.

In this way, they managed to save the original assumption of directional dependence and solve the problem. However, it brought about a new problem as to how the lengths could be measured to be the same, but as soon as light was traveling along those paths the lengths were shortened. Of course, this is not a rational solution.

The Doppler-Fizeau effect of the shifting of wavelengths of stars was well known at the time, and should have provided a clue to their mistake, as it was also widely thought that light traveled through the luminiferous aether. Maxwell's aether drag hypothesis also implies that the energy and wavelength of light has to have a directional dependence. If Fitzgerald and Lorentz had really thought about it, they might have realized that the length contraction fix is unnecessary, and doing away with Maxwell's false assumption of directional dependence solves the Michelson-Morely problem. **It is a lie to say that a length contraction hypothesis is required to explain the Michelson-Morley results.**

Even though length contraction is not necessary to solve the Michelson-Morley result, relativistic distance measurement shifts do occur when measuring the distance objects have traveled in a moving reference frame. But, does that mean that rod length contraction occurs?

Everyone should agree that a rod would have a standard length when it is observed in the aether rest frame. The rod can then be accelerated to a fixed velocity relative to the aether rest frame in the direction of the rod's length. Under the length contraction theory, the rod will now be shorter than it was when viewed by an observer in the aether rest frame. To an observer who is accelerated along with the rod and is still in the rod's frame of reference, the rod's length never changed.

At the same time, under the length contraction theory, the rod's width is the same for both observers. As with

Maxwell's theory, the length contraction theory still has a directional dependence, a dependence that does not exist.

Since a long physical rod cannot be accelerated to close to the speed of light, it has not been possible to perform a direct experimental test of the theory. Physical length contraction is unproven. There are some experiments that are thought by some to confirm length contraction, but they can be accounted for by normal interactions involving aether. As we have seen in many other instances, aether deniers have difficulty understanding what is physically going on.

We can also consider what would happen to a physical rod if it were contracted in a single direction. If the rod was formed from a salt or alloy that had a known crystalline lattice spacing that would mean that the crystal lattice was changed in one direction, but not in the other. The material would have to be denser in one direction than in the other. This would require a change in the underlying material physics.

Another thing to think about is how the rod interacts with the aether. Does the presence of quantum fluctuations, both around the rod and within the rod's structure, somehow change the rod's dimensions when the rod is moving relative to the aether? Since the rod is made of charged particles, the aether would rotate in response to the moving rod, but aether rotation alone would not change the dimensions. There is no interaction between the rod and the aether than could change the rod's external dimensions.

The length contraction believers do not have a model for a physical mechanism that could account for length contraction, whether they are aether deniers or not. In either case, length contraction is another magical theory unsubstantiated by actual physics.

Above and beyond this information, we also can consider how the change in observed distance traveled occurs, so we understand how distance traveled compares to length contraction. As mentioned in previous chapters, clock rates change in a reference frame moving relative to the aether rest frame. Moving clocks run slower than aether rest frame clocks. The reason that distances appear shorter to an observer in the aether rest frame is because that observer's clock runs faster, not because units of distance are shorter. A moving observer does not change physical distances in the aether rest frame.

Early relativity theorists came up with two possible explanations for changes in distance traveled, length contraction, and time dilation, and they came up with the wrong one first. Time t' does not equal t. Consequently x' does not equal x either, and to the observer in the aether rest frame, the distance traveled in the moving rest frame is shorter. It has nothing to do with actual physical distances or rod lengths. There is no mystery to it. In fact, if you try to apply both corrections simultaneously, as relativity theory suggests, you get the wrong result.

Length contraction is a lie. It has never been observed experimentally and there is no scientific explanation for how rod shortening could occur. The observed, real shortening of observed distance traveled due to clock rate differences was falsely applied to objects, when the objects in a moving reference frame do not experience any real contraction. A person with the faster clock just sees a shorter distance being traveled.

Lie #15: Space Contraction

The 15th of the greatest lies in physics is space contraction. Space contraction, like length contraction (Lie #14), is a phenomenon related to special relativity theory (Lie #13). Notice the pattern? With length contraction, an observer in the aether rest frame, observing objects in a second reference frame moving at a velocity that is significant with respect to the speed of light, sees a shortening of the distanced traveled. This creates a problem for believers of the false assumption that the speed of light is the same for all observers.

In order to save his false assumption, Einstein came up with the idea that when the length contracted, space was contracted and light was actually moving at the speed of light because the dimensions of space had changed. According to Einstein, the observer was just tricked into thinking that light was moving a shorter distance by appearing to move slower than the speed of light in free space.

This 'logic' runs into problems as soon as you have a second observer watching the same space. Let's say for example that you have a million observers moving at a million different velocities relative to the same object moving at a speed at or near the speed of light relative to all of them. If space contraction were correct, space would have to be contracted a million different amounts simultaneously. Distances in the aether rest frame do not change when an observer moves.

Space contraction is inconsistent with classical geometry, but having two or a million different geometries simultaneously in the same space is ridiculous. Any person with critical thinking ability should instantly recognize it as a fallacy.

The space contraction hypothesis also ignores that the quantum field has energy, as Einstein, the most famous of quantum fluctuation deniers, proposed it. The most energetic quantum fluctuations are those with the shortest wavelengths. In theory the wavelengths have all possible lengths, leading to a quantum field with an infinite amount of energy over any volume of space. Traditionally the Planck Length (~10^{-35} meters) is used as the smallest possible wavelength, or cut-off wavelength, since we are unsure if the normal laws of physics hold at distances that short and shorter.

If we use the Planck Length as a cut-off wavelength, the zero-point energy is equivalent to ~10^{95} grams per cubic centimeter. John Wheeler and Charles Misner originally published it as 10^{94} grams per cubic centimeter,[33] but an order of magnitude or so at this energy is not terribly important. Based on current estimates of 10^{56} grams in the visible universe under the big bang model, every cubic centimeter of vacuum has much more energy than the entire visible universe.

Quantum fluctuations have wavelengths and frequencies. It is their wavelengths that establish the physical dimensions of space. The physical dimensions of the aether rest frame are uniform in all directions in free space. That said, the presence of mass increases the quantum van der Waals torque of the quantum field. This increase the local permittivity and permeability of the quantum field reduces the speed of light. This phenomenon leads to effects falsely lumped together with general relativistic effects. But space still does not have dimensions on its own and the presence of matter does not change that.

The idea that a photon could somehow compress that much zero-point energy, and change the dimensions of space by changing the wavelengths of all those quantum fluctuations is ridiculous. A photon simply does not have enough energy, nor a form of interaction, to cause space to contract.

The reason for the space contraction lie is the same as the length contraction lie. An observer in the aether rest frame watching light move in a moving frame of reference, truly does see light move a shorter distance in the moving frame, while moving the correct distance as measured in his frame of reference. An observer in the moving frame of reference will see light move the correct distance because his or her clock runs slow. This has nothing to do with length or space contraction.

As before, there were two possible explanations that physicists could come up with. The first they came up with was length contraction, which was later extended to include space contraction. The second possibility that came a short time later was clock slowing. When clock slowing appeared, physicists were already coming to accept the length contraction model and failed to recognize that clock slowing by itself could be seen as the solution. They also failed to recognize that when you apply both corrections you get the wrong result.

It is impossible to stress enough that the aether rest frame is the standard rest frame. Without a standard rest frame, it becomes impossible to determine which observer is stationary and which observer is moving. Without a standard reference frame, you also have the problem of whose clock is faster and whose is slower, the underlying problem with the twin paradox. If there were such a thing as space contraction, we would also not know which observer's space was being contracted. It is a good thing that there is a standard frame of reference, the aether rest frame. That way we have a standard to compare two frames of reference and a standard ruler for measuring distances.

Space contraction is a lie. The speed of light only appears to be the same for an observer in a moving rest frame because their clocks are slower. Distance units in space do not change, because quantum fluctuation wavelengths in free space cannot be changed.

[33] J. A. Wheeler and C. Misner, Geometrodynamics, Academic Press, New York, 1962.

Lie #16: Time Dilation of Space

The 16th of the greatest lies in physics is time dilation of space. This is not to say that relativistic clock slowing does not exist. Clock slowing does exist and has been proven experimentally, most notably with Global Positioning System (GPS) clocks in orbit. While many people use the terms 'time dilation' and 'clock slowing' interchangeably, there is an important distinction, and that distinction tells us that time dilation is a lie.

Predictably the trouble comes back to aether rejection. With Einstein the aether denier inventing special relativity to replace Lorentzian relativity, he lacked a physical explanation for clock slowing. In a universe without quantum fluctuations, clock slowing had to be a function of time being slowed due to space. It implied that space itself has clocks that slow when they move, and causes other clocks, both mechanical and biological to slow. The concept of time dilation as a property of space is what makes time dilation a lie. Space by itself is non-physical and does not contain physical clocks.

In special relativity theory, the time dilation concept has an additional problem because there is no preferred frame of reference. If you have two frames of reference with observers in each, both observers think they are stationary and the other one is moving. How does space know when it is supposed to have time dilation? Of course, there is no way to resolve this problem without a universal standard reference frame, the aether rest frame.

In Lorentzian relativity, aether does exist, and so his theory is in line with the physical evidence for the existence of aether. In this case, clock slowing—mechanical or biological—is due to interactions with the aether. Clock slowing is a physical phenomenon that is explained by physical interactions with the aether. Clock slowing is not a magical property of space.

Understanding the physical mechanism is not that challenging if we start with the simplest physical clock, a quantum fluctuation dipole. We can consider a dipole like those in a photon. Photon dipoles rotate 180 degrees during their life and their wavelength and frequency are consistent with the speed of light, **λv = c**. These dipoles have their longest wavelength when viewed by an observer in the aether rest frame. At the same time their frequency, their clock as it were, is fastest in the aether rest frame

To an observer in a moving frame of reference, the dipole is moving against the motion of the aether dipoles in the aether rest frame. The aether rest frame dipoles appear to exert a real physical resistance to the motion of our dipole-clock inducing a van der Waals torque. The dipole-clock appears to have a shorter wavelength (i.e. distance traveled) to a stationary observer. At the same time the stationary observer sees the dipole's rate of rotation appear to slow when he or she views it from the perspective of a moving observer. As the dipole slows, its frequency is reduced, and so its clock runs at a slower rate.

The result from multiplying the wavelength and frequency is the same for both observers, so they observe the same speed of light. They do, however, see the light as a different wavelength. To the moving observer the light is redshifted. The clock slowing experienced by the quantum dipole is due to the physical interaction with the aether. Or, anther way to think of it is that the dipole behaves as if it is always in the aether rest frame, while the moving observer's clock is slowed due to the clock's motion against the aether.

Time dilation of space is a lie. **It is a lie to say that empty space even has a clock.** The clock is the aether, and clock slowing is due to interactions with the aether.

Lie #17: Action at a Distance

> *That one body may act upon another at a distance through a vacuum, without the mediation of any thing else, by and through which their action and force may be conveyed from one to another, is to me so great an absurdity, that I believe no man, who has in philosophical matters a competent faculty for thinking, can ever fall into it.*[34]
>
> Sir Isaac Newton, 1693

The 17th greatest lie in physics is action at a distance. Action at a distance is another of the many great lies born from the minds of quantum fluctuation deniers. Deniers think space is empty so there is nothing available to them to be a medium for force transmission. That leaves them with a theory of magic, where forces are transported magically from one place to the other. Newton was smart enough, unlike most modern physicists, to realize that magical force transmission, or action at a distance, was not the answer.

Newton also cannot be blamed for choosing not to speculate about how forces are transmitted, as there was no convincing aether or quantum fluctuation theory during his day. They did have variations on Descartes old vortex theory, and Nicolas Fatio de Duillier came out with a theory later popularized by Georges-Louis Le Sage involving corpuscles.[35] But both of those theories had problems and Newton wisely chose not to champion them too strongly. Once Max Planck described quantum harmonic oscillators leading to the concept of quantum fluctuations, modern physicists did not have an excuse.

Action at a distance theory leads to so many questions, such as:
1. How does a signal jump from one location to another?
2. How is this signal sent such that the signaling body does not lose energy?

3. What sort of signal requires no energy?
4. How is data stored and transmitted?
5. How does the signal travel through space?
6. How fast does it travel?
7. How does the receiving body receive the signal?
8. How does the receiving body process a large amount of data from numerous signals?
9. How does the receiving body move in response to the signals?
10. What is the source of the energy behind motion?
11. What is the physical mechanism responsible for motion?

As Newton stated, surely no one with a capacity for critical thinking should believe in magical action at a distance. There are far too many unsolvable problems to solve when one denies the existence of a transmission medium.

Fortunately, quantum fluctuations do exist, so there is a medium throughout space that transmits forces and causes objects to move. We even have a model for how forces are transmitted through the vacuum as discovered by Casimir. He made the scientific leap in recognizing that there is quantum field pressure derived from van der Waals forces between quantum fluctuation dipoles. This Casimir-van der Waals force pushes on all objects, but the force is normally the same in all directions and is therefore not detected unless the pressure becomes greater or smaller in the region between bodies of matter.

Action at a distance is a lie. The existence of aether and the existence of Casimir-van der Waals forces tells us that there is no need for magical action at a distance.

[34] I. Newton, excerpt from a letter to Dr. Bentley dated 25 February 1693.
[35] N Fatio de Dullier, "De le Cause de la Pesanteur" (ca 1690), Edited version published by K. Bopp, Drei Untersuchungen zur Geschichte der Mathematik, Walter de Gruyter & Co. pg 19-26 1929.

Lie #18: Gauge Bosons

The 18th of the greatest lies in physics is gauge bosons. Gauge bosons are the quantum fluctuation deniers' answer to the fallacy of action at a distance, without actually solving any of its problems. In gauge boson theory, forces are mediated through the exchange of particles, the gauge bosons. These particles hypothetically bridge the gap in space so that instructions may pass from one object to another.

Each type of force and its associated field is said to have a specific type of gauge boson associated with it. There are photons for electromagnetic forces, gravitons for gravity, gluons for strong forces, and W and Z bosons for weak interactions. Lie #7 covers some reasons why photons cannot be gauge bosons.

The first point of trouble comes with conservation of energy, as there is no energy lost by either body during this hypothetical particle exchange. The principle of conservation of energy cannot be violated. Consequently, gauge bosons must be virtual particle pairs, or as they are also referred to here, quantum fluctuations. As quantum fluctuations, gauge bosons must be consistent with Planck's theory of the quantum harmonic oscillator and cannot exceed the energy-time constraints of Heisenberg's uncertainty principle. If gauge bosons exceed the Planck oscillator energy, they violate the principle of conservation of energy.

Most physicists who support gauge boson theory conveniently leave out the requirement that quantum fluctuations be composed of virtual particle pairs. Each pair must have a matter and antimatter particle such that the balance of matter versus antimatter is conserved. Each pair must also have opposing charges so that electrical neutrality is observed. Or each particle must be electrically neutral, assuming there are any real elementary particles that are neutral.

Additionally, properties such as momentum, angular momentum, magnetic moment, and spin must be conserved. These properties cannot come from nothing. They also cannot be taken away from a particle without changing that particle.

Gauge boson theories treat gauge bosons as individual, supposedly virtual, particles. It is also said they are emitted by one particle and absorbed by another, and somehow the properties of the emitting and receiving particle remain unchanged. If that were the case, a number of types of energy, momentum, and other particle properties would not be conserved.

There is also the thorny issue of mass. A single particle that normally has mass in a free state—that lives long enough to exceed the Planck energy—must have mass. This limits the distance that a hypothetical virtual single particle can travel. With certain heavy gauge bosons, they cannot be thought to have the necessary range needed to fulfill their gauge boson function without violating the principle of conservation of energy. It is pretty easy to see the flaws in the logic.

If quantum particle pairs could act as gauge bosons, they would need to have a nearly infinite range of wavelengths and energy. A quantum fluctuation of a given energy has one wavelength and one frequency as those properties are all fixed by a single formula. Consequently, a quantum particle pair stretched between two bodies can only carry a single point of data, the distance between the two bodies.

As mentioned before, the direction and energy of a quantum fluctuation cannot be measured simultaneously per Heisenberg's uncertainty principle. An absorbing particle could not tell where a hypothetical gauge boson came from, if it knows the energy. And if it can determine where it came from, the energy is unknown. So, even though a virtual gauge boson can have energy and direction, it can still only be used to transmit a single bit of data; otherwise, the uncertainty principle would be violated.

A gauge boson therefore cannot carry information of a particle's, mass, charge, physical dimensions, or any other information needed to determine the magnitude and direction of the force supposedly being transmitted.

The gauge boson hypothesis fails to describe:
1. How are gauge bosons generated?
2. How are energy, momentum, angular momentum, matter-antimatter, electric charge, magnetic moment, or spin conserved?
3. How do gauge bosons with mass travel farther than allowed by the conservation of energy principle or speed of light limit?
4. Where do gauge bosons store their data?
5. How do gauge bosons travel without using any energy?
6. How does the second particle know a gauge boson is there?
7. How does the second particle receive the gauge boson?
8. How does the receiving particle know how far the gauge boson traveled and from what direction?
9. How does the receiving particle measure, or download the gauge boson's data?
10. How would the receiving particle compute the result?
11. What sort of supercomputer is used to integrate the data from billions of gauge bosons received simultaneously?
12. How does the receiving particle correct for speed of light delays, since a gauge boson can only transmit at the speed of light?
13. Once a computation is complete, how does the body move?
14. Where does the energy come from to make the body move?

The gauge boson hypothesis is essentially no better than action at a distance. It fails many of the same tests as is obvious when the lists from the past two chapters are compared, and they are by no means intended to be exhaustive. The gauge boson theory fails to yield a solution to most of the same basic problems as action at a distance. It only changes the nature of the problems.

The gauge boson model is a lie. There has to be a continuum of quantum fluctuations, the aether, between two bodies to mediate a force. To paraphrase Newton, no one with a capacity for critical thinking should believe in gauge bosons.

Lie #19: Mass is Intrinsic

The 19th of the greatest lies in physics is mass is intrinsic. Mass is a fundamental property of all particles and materials. Mass is a measurement of the amount of material. Mass is related to weight, as weight on Earth is mass times the acceleration due to gravity, and all mass is ultimately measured by accelerating it. The mass of any body is the sum of the masses of its constituent particles, most importantly, the most stable particles, electrons, protons, and neutrons. All of the stable matter in the universe is composed of those three particles.

When we consider the mass of a particle, we run into problems. The standard model of physics does not include a physical explanation of mass. Physicists can measure the mass of particles including the electron, proton, and neutron, but they do not understand how particles have mass. In the absence of an explanation, physicists treat mass as an intrinsic property of a particle. Mass just is. Even in Dirac's equation, mass is simply a number that has not been derived from something more fundamental.

The three most stable particles are only the tip of an iceberg-sized problem. Based on scientific experiment, and within the standard model, there are 3 types of electrons with mass, 3 types of neutrinos that may have mass, 6 types of quarks with mass, 3 gauge bosons with mass, 32 mesons with mass, and 75 baryons with mass.[37] That is as many as 122 particles with mass, and the standard model does not contain a theory that explains any of them. And, there are many more particles that have been detected that do not fit the quark model, and many predicted by the quark model that have not been detected. There is no way there are 122+ intrinsic constants in a fundamental theory of anything.

Treating mass as intrinsic is not scientifically sound; in fact, it goes against the very definitions of science and physics. Science requires that we study the structure and behavior of things and physics requires that we be able to physically and mathematically explain that structure and behavior. Calling something intrinsic is an act of negligence; it is scientists failing to do their jobs.

As with so many other lies, this one, at least in part, is due to physicists ignoring aether. We do know that mass is related to energy by the well-known equation **E = mc²**, where **c** is the speed of light and **E** and **m** are energy and mass respectively. That equation means that we should be able to understand and physically explain mass in terms of energy. Aether deniers, however, who consider an electron or proton to be in completely empty space, are stuck. There is no way to model mass as energy if there is no form of energy in and around the particle.

Fortunately, we know that quantum fluctuations do exist as zero-point energy and form the zero-point field, otherwise known as aether. Even when the Planck length is used as a cutoff wavelength, the vacuum has energy equivalent to $\sim 10^{95}$ grams per cubic centimeter. That opens up the possibility that mass can be modeled with respect to the energy of aether. Frankly that is the only physically real option we have.

It is also fortunate that one of the most brilliant physicists of the last century, Dirac, already came up with the idea for an explanation for mass. Dirac thought that a particle's energy may be due to the energy it takes for it to maintain its place in the Dirac Sea. The Dirac Sea was his hypothetical form of aether composed of electrons and positrons. Aether pushes against particles and particles must have a certain amount of energy to push back and maintain their place in the Dirac Sea. He hypothesized that the 'push back' energy is the particle's mass. Sadly, it appears that neither he nor anyone else bothered to perform the calculation to see if his model was correct until I did in 2011.

The simplest way to put Dirac's model into a form that is easy to calculate is to assume that the particle is like a spherical shell. Then we can recognize that quantum fluctuation wavelengths equal to any of the range of diameters of that shell cannot exist in or around it. Those wavelengths are excluded in a similar way that wavelengths are excluded between two parallel plates in the Casimir effect. Essentially, we can treat a shell-like model of a particle as a Casimir cavity and measure the energy.

The proton has a known charge radius so that is a good starting point for it. The electron on the other hand is a bit more confusing as when we scatter light or electrons off an electron it appears to be have physical size related to Compton scattering. Also, the mass of the electron equates to the Compton wavelength which is much larger than a proton. On the other hand, if we scatter high-energy protons off an electron it appears to be much smaller than the proton, and perhaps even a point. The difficulties with the point mass hypothesis are addressed in the next chapter.

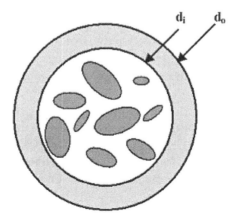

Figure 19-1 An illustration of the particle shell with elliptically shaped quantum fluctuations in the middle and the circles describing the inner and outer spherical shells with diameters indicated.

The calculation shows that the mass of a proton is equal to the quantum energy excluded by a shell with the proton's charge radius. The mass of the electron is equal to the quantum energy excluded by a shell with a diameter equal to its Compton wavelength. Note that the charge radius of the electron has never been formally determined at that scale through scattering experiments. The ratio between the masses, one of the great mysteries of physics, is simply a function of the quantum field energy and the particle radii. I explain this in greater detail in my paper "Proton and electron mass derived as the vacuum energy displaced by a Casimir cavity."[38]

It turns out that Dirac was correct in his hypothetical particle model for mass-energy, but it is sad that it took 80 years for someone to perform the computation to prove it. Since the neutron is similar in size to a proton, its mass is explained in a similar fashion.

The masses of the three main particles that make up the stable mass in the universe are explained very simply with a model originally described by Dirac. Mass is displaced zero-point energy. Admittedly there must be a different type of zero-point energy displacement to describe the masses of the unstable 'particles,' the resonances. Their mass turns out to be relativistically increased mass of their component electrons and/or protons. That will be discussed more later.

The concept of intrinsic mass is a lie as it must be possible for all particle masses to be explained in both a physical and mathematical way. The science of physics requires it. **It is also a lie that proton, electron, or neutron masses are fundamental constants.**

---

[37] Per Wikipedia's lists of mesons and baryons.
[38] R. Fleming, " Proton and electron mass derived as the vacuum energy displaced by a Casimir cavity," https://www.researchgate.net/publication/283795493_Proton_and_electron_mass_derived_as_the_vacuum_energy_displaced_by_a_Casimir_cavity, 2012.

Lie #20: Point Mass

The 20th of the greatest lies in physics is point mass. In standard model physics, physicists have basically given up on attempting to describe particles as physical bodies, and have taken to treating particles as points, in other words with no physical dimensions. The point mass hypothesis is so absurd it deserves its place among the greatest lies in physics.

The first difficulty is density. If we first consider a particle that takes up a small volume and we consider mass to be distributed within that volume, then we can calculate a density, mass per unit volume. As this hypothetical particle gets smaller, the density increases. If we consider a point particle then the density is infinite and mass density cannot be infinite.

Another difficulty is the number of particles as addressed in the last chapter. When we have a point particle, we are left with an intrinsic mass theory. But we have 122 particles and resonances and perhaps many more, and every one of those 'particles' has a different mass. There is no fundamental theory of anything that has 122+ intrinsic masses for 122+ separate particles. Keep in mind that even when these 'particles' are treated as composite particles, the current model, the quark model, does not address the mass problem.

Here it is important to note that having a one-dimensional model, such as in string theory, does not help. We still have 122+ different one-dimensional strings, and the only way to resolve them is by having different masses per unit length. But the mass density per unit volume is still infinite and irrational. The only way to escape the irrational infinity problem is to have a three-dimensional particle model.

Next, we must consider that if we model the masses of particles, that model must involve quantum field energy, since zero-point energy does exist and ultimately is the fundamental source of all energy. In order to have a particle mass description, the particles must have a volume, and if they have volume, they are displacing zero-point energy. And, as Dirac surmised, the displaced energy is particle mass-energy.

So, if an electron, for example, has a classical radius spherical shell of 10^{-15} meters average diameter, it will have a mass only slightly smaller than that of a proton. If it has a shell 10^{-18} meters in diameter it will be 1000 times more massive than a proton, and at 10^{-21} meters a million times for massive. If it is the size of the Plank length, $\sim 10^{-35}$ meters, it will be a small black hole.

Keep in mind that if we model a particle as a solid sphere instead of as a shell, the mass would be infinite, as even the smallest quantum fluctuations would be excluded. The same problem occurs with a spherical shell that is infinitely smooth, as small quantum fluctuations within the shell would also be excluded. The shell structure of a particle, whatever it is, must be porous to smaller quantum fluctuations and particles.

The concept of a point mass is a lie. It impossible to model the mass of a point particle as it would be infinitely dense and infinitely massive. More realistically it would have no mass. Particles must have real physical volumes.

Lie #21: Matter and Antimatter are Intrinsic

> *My own opinion is that we ought to search for a way of making fundamental changes not only in our present Quantum Mechanics, but actually in Classical Mechanics as well. Since Classical Mechanics and Quantum Mechanics are closely connected, I believe we may still learn from a further study of Classical Mechanics. In this point of view I differ from some theoretical physicists, in particular Bohr and Pauli.*[39]
>
> Paul Dirac, 1949

The 21st of the greatest lies in physics is the concept that matter and antimatter are intrinsic properties. According to Dirac's interpretation of his solutions to Dirac's equation, matter and antimatter have a positive and negative 'energy' relationship. Most mainstream physicists, however, do not attempt to understand this 'energy' relationship, and simply treat matter and antimatter as intrinsic properties. While this may seem insignificant on the surface, it is a huge mistake, perhaps second only to the denial of aether.

Dirac's equation of the electron was one of the great triumphs of physics of the 20th century. It was puzzling, however, that it had both a positive and a negative 'energy' solution, and there was no known condition that prohibited the negative 'energy' solution. Dirac interpreted them as negative and positive 'energy,' **mc²** and **-mc²**, but no one knew for certain what **-mc²** meant. In any case the negative 'energy' solution is now accepted as a prediction of the positron, although Dirac admitted he was not brave enough to publish that prediction. Physics historians have been kind to him and deservedly given him the credit anyway.

The negative 'energy' solution was and still is puzzling, as all forms of real energy are positive. Dirac got around the problem by imagining the vacuum is composed of a sea of virtual positive and negative electrons, the so-called Dirac Sea. As we have seen, his model was not far from the truth as we have come to know it. His Dirac Sea model was an early model for the zero-point field and aether and it could have led to many early advances in physics if it had been taken seriously.

Dirac then saw the electron and positron as holes in the Dirac Sea, which leads to a simple mathematical derivation of mass described in chapter #19. While Dirac had explained mass-energy, he had not actually successfully addressed the positive and negative, matter and antimatter 'energy' solutions to his equation.

Dirac did attempt to model the positive and negative 'energy' solutions to his equation using his Dirac Sea model. He considered that there could be a positive energy space above a negative energy space separated by a region of zero energy, with electrons existing above the neutral zone and positrons below. Matter and antimatter could then be opposite but equal in terms of something we can think of as matter-energy and can cancel each other out through annihilation.

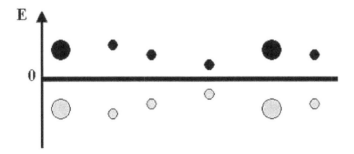

Figure 21-1 An illustration of the Dirac Sea, where pairs of virtual particles exist on opposite sides of a line indicating zero energy. The black circles are positive and the gray circles are negative.

Alternatively, he also envisioned matter as a bubble in space and antimatter as a hole, such that they are equal but opposite in some respect. In this model, a bubble could fill a hole and turn that space into normal vacuum. In this model, a hole and bubble neutralize each other in an annihilation process.

That is where Dirac's ideas stalled out. The aether deniers, while willing to listen to Dirac due to his brilliance, were not persuaded that his aether-like Dirac Sea could be real. His bubble and hole models where considered to be mere curiosities, although they were still passed down to generations of new physicists. And it turns out that actual holes in solid state materials can act like particles, much in the way that Dirac imagined.

Physicists failed to spend time and energy trying to understand the meaning of the positive and negative 'energy' solutions to the Dirac equation. To most physicists, matter and antimatter became intrinsic properties of particles. This amounted to a denial of the importance of the matter-antimatter 'energy' relationship. To this day matter and antimatter are presented as intrinsic properties. That is a lie. There must be more to it.

To understand it we need to first recall that while Dirac described the positive and negative solutions as 'energy,' they do not represent energy as it is usually defined. This new type of 'energy' relating to matter and antimatter should perhaps be referred to as matter-energy to keep it separate.

There is another particle property that has exactly the same type of quasi-energy relationship with two states in opposition. That property is electric charge which can be thought of in terms of charge-energy, and modeled in the same way that Dirac conceptually modeled the matter-energy relationship. When Maxwell was puzzling over electrical interactions, he sometimes described charge as a sort of positive and negative charge-energy, not too dissimilar from the way Dirac described matter-energy.

Perhaps electric charge can even be modeled in a similar mathematical fashion to matter and antimatter in Dirac's equation. Matter and antimatter have the same type of relationship as positive and negative electric charge, while being a separate but related particle property.

The statement that matter and antimatter are intrinsic is a lie. Physicists missed out on an important discovery when they chose to disregard the implications of negative solutions of the Dirac equation. There is a physical explanation for matter hidden in the interpretation of the Dirac equation and the physical structure of the electron. There is still much to learn about positive and negative matter-energy interactions, as we will see.

[39] P.A.M. Dirac, "The Relation of Classical to Quantum Mechanics." Proceedings of the Second Mathematical Congress, Vancouver, 1949. Toronto: University of Toronto Press. Pg. 18 (1951).

Lie #22 Inertia is Intrinsic

> *For whether there be any intrinsically material inertia or not, there certainly is an electrical inertia. [...] Quite possibly there is no other kind. Quite possibly that which we observe as the inertia of ordinary matter is simply the electric inertia, or self-induction, of an immense number of ionic charges, or electric atoms, or electrons. This is by far the most interesting hypothesis, because it enables us to progress, and is definite. The admixture of properties – partly explained, viz. the electrical, partly unexplained, viz. the material – lands us nowhere.*[40]
>
> Oliver Lodge, 1906

The 22nd of the greatest lies in physics is inertia is intrinsic. Inertia is the nature of a body to maintain its velocity unless acted upon. As with many of the great lies of physics, quantum fluctuation deniers are behind this one too. Without quantum fluctuations forming a zero-point field or aether, there is nothing in their vacuum model capable of interacting with a body to cause inertia. The deniers are then stuck with inertia happening as if by magic, intrinsically, completely unexplainable by them.

Worse than that, quantum fluctuation deniers refuse to ask the question and belittle those who do. This is very sad, and non-scientific, as inertia is the cornerstone to mechanics and mechanical force theory. There is no theory of everything without a physical explanation for inertia.

It turns out that it is not all that hard to understand inertia when one accepts that quantum fluctuations exist, and consider their make-up and interactions. The vacuum is filled with quantum fluctuations, and each one can interact with nearby bodies of matter. The question is how? To answer that, we need to go back to the description of the quantum fluctuations.

One of the main reasons that previous attempts to describe inertia as an interaction with quantum fluctuations failed, is that they were modeled as photon pairs. As addressed in Lie #5, the zero-point field cannot be composed of photon pairs, as any physically real model of a virtual photon pair exceeds the energy of a quantum harmonic oscillator, and would thus exceed the Planck energy, Heisenberg's uncertainty principle, and violate the principle of conservation of energy.

Quantum fluctuations are real particle pairs, such as electron-positron and proton-antiproton pairs. These particle pairs form electric charge dipoles, and they form what we can think of as matter-antimatter dipoles. If we go back to Dirac's equation of the electron, electrons are a positive 'energy' solution and positrons are a negative 'energy' solution. Matter and antimatter are an equal but opposite form of 'energy' and cancel each other out. They truly do form a dipole of sorts that is distinctly different from the electric charge dipole.

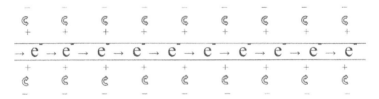

Figure 22-1 Electrons in motion within a wire induce rotation in surrounding quantum fluctuation dipoles. The rotation of the quantum fluctuation dipoles induces a current in the wire.

There is a phenomenon from electromagnetic theory that is very similar, if not identical to inertia. The phenomenon is self-induction. If we consider a current flowing in a closed circuit, the current induces a magnetic field in the aether around the wire. The magnetic field around the wire also induces a current on the wire. In a hypothetical lossless circuit, this self-induction could continue indefinitely, in the same way that a moving body in a quantum vacuum, in otherwise empty space, could continue indefinitely.

While it is true that otherwise electrically neutral bodies of matter are composed of electrically charged particles when viewed at a subatomic scale, the body still appears neutral at any distance, and does not produce a magnetic field when it moves. Inertia cannot be ordinary electromagnetic self-induction.

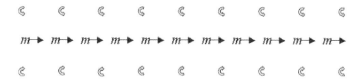

Figure 22-2 Matter in motion along a line inducing rotation in surrounding quantum fluctuations.

Bodies of matter are, however, matter, and as such can interact with Dirac's matter-antimatter dipole, assuming Dirac's 'energy' interpretation is valid. If true then inertia is simply self-induction within the scope of a matter-antimatter interaction. If the dipole is not matter and antimatter, a strong case can be made that there must be a non-electric dipole of some kind in order to explain inertia and other mechanical forces. But, if one looks at all the properties of an electron-positron pair, there are only two dipoles with two opposite types of charge or 'energy.' So the simplest of all our options is that inertia is a type of matter-antimatter self-induction.

The statement that inertia is intrinsic is a lie. With or without the matter-antimatter dipole explanation for inertia, it is clear that the idea that inertia is intrinsic to a body is false. We must consider inertia as a body interacting with the zero-point field in a manner consistent with self-induction.

[40] O. Lodge, Electrons or The Nature and Properties of Negative Electricity, George Bell and Sons, London, 1906.

Lie #23: There is no Mechanical Force

The 23rd of the greatest lies in physics is there is no mechanical force. If you look at the four fundamental forces in the standard model, electromagnetism, gravity, weak, and strong, there is no accounting for mechanical forces. The failure to include mechanical forces in the list of fundamental forces is one of the great blunders of standard model physics, as the fundamental forces cannot ultimately be unified without including mechanical forces.

Newton's second law, **F** = **ma**, force equals mass times acceleration, gives us the simplest mathematical representation of a mechanical force after inertia. This unit of force is called the Newton in his honor. The most common mechanical force we deal with every day is the weight of an object, where the acceleration is the acceleration due to gravity.

Mechanical forces were the earliest forces to be described mathematically. That is why it is so baffling that mechanical forces do not get mentioned among the so-called fundamental forces. The most likely culprit is that many physicists consider mechanical forces to be a closed theory. They think they already know everything they need to know about mechanics. As we have seen, this is not the case. They do not even understand inertia, the most basic mechanical force principle.

The other failing is that unlike electricity and magnetism, mechanics never really had a Michael Faraday moment, someone who stood up and said, aha, there must be some kind of field to explain the forces. As was discussed in Lie #22, any rational discussion of a model for inertia leads us to the existence of a field.

There must be a field to explain inertia. When a body moves through space it produces a field in the aether, and then the field in the aether makes the body continue to move. It is a form of mechanical self-induction. Without some kind of field, we have no hope to successfully model inertia. Also, without a mechanical field, there is no hope of unifying electromagnetic and mechanical force theory. It is important to note that the speed of light limit is the same for electrically neutral and charged objects.

The missing mechanical field is easy to recognize when studying a spinning top or gyroscope. Perhaps that is why they amaze us so. As a toy top spins on its tip, it starts to fall, accelerated by gravity. This acceleration is, however, translated into a motion perpendicular to gravity, in the direction of the top's precession, which is in the same direction as the top's direction of rotation.

Then the acceleration in the direction of precession is translated into an upward acceleration that opposes gravity. In mechanics, in accordance with Newton's Second Law, if there is acceleration there is a force. A spinning toy top produces mechanical forces opposing gravity. Physicists like to call it a torque and pretend that it is less than a true force, a so-called pseudo-force.

The top must, however, be pushing against something. And, since mechanical forces exist in a vacuum, the top must in some fashion grab hold of the vacuum, or specifically the quantum fluctuations. There must be a field, and the field must be composed of quantum fluctuations. A top even obeys rules, like those from electricity and magnetism, including having similar types of fields with polarization and rotation of quantum dipoles. The typical toy top, of course, does not have a net electric charge, and does not produce electromagnetic fields, so there must be another type of dipole and non-electric fields.

Since the late 19th century many physicists, including notably Oliver Heaviside, Robert Forward, and Martin Tajmar, recognized that mechanical force theory could be described in a manner similar to electricity and magnetism.[41,42,43] They each attempted to fashion a theory by uniting the mechanical force with gravity into a gravitomagnetic theory.

Those theories fail in a large part due to gravity working in the wrong direction. Unlike electric charges, masses in the gravitational case are attracted rather than repelled. Those theories also lack a charge-like dipole such that opposite charges attract and like charges repel. There is no negative mass or negative gravity.

If there were a force in the correct direction for an electromagnetic-like mechanical force theory to work, that would mean the universe would be expanding at an accelerating rate. Oh wait, it is.

It would also mean that stars traveling in the same direction side-by-side would be attracted to each other, and in rotating galaxies, stars would be affected by a mechanical Lorentz-like force pushing stars with the highest tangential velocities toward the galactic core. These two effects would force rotating galaxies to form spirals, so we know that is wrong, oops, well, certainly physicists should not have missed anything that obvious. It is funny how if there is a mechanical field, and associated dipole, all the major unsolved problems in mechanics are readily fixed. Understanding inertia as a form of self-induction interaction with the quantum field leads to an electrically neutral force that obeys Maxwell's equations.

The underlying problems were that physicists were not aware of a charge-like dipole to go with their mechanical force theory, and most physicists failed to grasp that mechanics is a field force. There have been attempts to explain how electric dipoles could account for field-like mechanical interactions. They have so far failed. So for now we can say that this mechanical field force is not due to electric dipoles.

When Dirac came up with his equation of the electron, which predicted both positive and negative 'energy' electrons, a new possibility opened up. The new particle became known as the positron and the form of negative 'energy' became known as antimatter. Physicists failed to recognize that these positive and negative 'energy' solutions representing matter and antimatter looked like a new and different type of charge, a new dipole.

Some physicists will want to propose a different dipole. They may look at the properties of an electron and positron in the table below and pick one, but there are not any other good candidates from among the known particle properties. I guess they could make up some new science fiction property instead, as they tend to do.

Property	Electron	Positron
Group	Lepton	Lepton
Charge	$-e$	$+e$
Spin	½	½
Mass	0.511 MeV/c^2	0.511 MeV/c^2
Magnetic Moment	$-1\ \mu_B$	$+1\ \mu_B$
Matter - Antimatter	Matter	Antimatter

Table 23-1 Comparison of known fundamental properties of an electron and positron.

The denial of the mechanical force is truly one of the greatest lies in physics. There most certainly is a mechanical force and it is necessary for us to describe numerous phenomena across the field of mechanics. In order for mechanical force theory to explain how bodies interact with the vacuum, and to be unified with electromagnetic force theory, mechanical forces must be treated as a field theory that obeys a form of Maxwell's equations. There must be a mechanical dipole. For more information please refer to my paper "The Electro-Matter Force" or my book *The Zero-Point Universe*.[44]

[41] O. Heaviside, "A gravitational and electromagnetic analogy" The Electrician 31: 81–82 1893
[42] R.L. Forward, "General Relativity for the Experimentalist," Proceedings of the IRE, 892-586, 1961.
[43] M. Tajmar et al, Coupling of Electromagnetism and Gravitation in the Weak Field Approximation, Physica C 385:551-554 2003.
[44] R. Fleming, "The Electro-Matter Force," researchgate.net, 2012.

Lie #24: Physics Explains Mechanical Motion

> *All of our experience, without a single exception, enforces the proposition that no body moves in any direction, or in any way, except when some other body in contact with it impresses its own motion upon it. [...] For mathematical purposes, it has sometimes been convenient to treat a problem as if one body could act upon another without any physical medium between them; but such a conception has no degree of rationality, and I know of no one who believes in that as a fact. If this be granted, then our philosophy agrees with our experience, and every body moves because it is pushed, and the mechanical antecedent of every kind of phenomenon is to be looked for in some adjacent body possessing energy; that is, the ability to push or produce pressure.*[45]
>
> Amos Emerson Dolbear, 1897

The 24th of the greatest lies in physics is physics explains mechanical motion. This is equivalent to the same problem in electromagnetic theory (Lie#9) but for mechanical motion. Physicists pretend that they know how tops and gyroscopes move because they can calculate a result, when they in fact do not understand how spinning objects interact with and push against the aether.

As with electromagnetic theory, most physicists consider mechanical theory closed, a complete theory, never to be looked into again. Never mind that they cannot explain something as simple as a spinning top. Consequently, they never got around to describing how bodies physically move in response to mechanical forces when there is no physical contact. Note that at the subatomic level even physical contact is not what it seems.

As mentioned previously, physicists never even recognized that mechanics has to be a field force, even less come up with a physical model for the fields. Then in terms of actual motion, they neglected to ask what does the pushing and how is that push generated? Or if they did ask, they were quickly hushed and told not to worry about such things; it is heresy to discuss it. Fortunately, quantum fluctuations do exist, they do interact as part of a field force, and as with electromagnetic forces, they make a perfect medium for transmitting mechanical forces between bodies, which are not otherwise in physical contact.

As with electromagnetic force motion, it was Casimir who, without realizing it, figured out the fundamentals of how mechanical forces work in response to quantum field interactions. Quantum fluctuation dipoles produce van der Waals forces, and those forces push on bodies. In mechanical forces we simply have a different dipole. Except for the dipole, the principles are exactly the same.

As with electric dipoles, the van der Waals pressure force due to the mechanical (matter-antimatter) dipole changes when some quantum fluctuations are excluded from a region of space. Note that the electric and mechanical dipoles must be in some way different such that mechanical motion does not produce electromagnetic fields. I will not speculate on that here, since that ultimately gets into questions of particle structure. In any case we will need to determine a magnitude for mechanical charge and derive or experimentally determine appropriate constants for the mechanical force equations.

As with electromagnetic forces, two bodies with opposite mechanical charge polarize the quantum fluctuations between them. Then as dipoles come together and annihilate, the dipoles next to them are brought closer together. This leads to a reduction in the van der Waals pressure between two bodies with opposite charges. Then the outer pressure pushes the bodies together.

When two like charges (e.g. matter or matter) exist in a region of space, the quantum dipoles at the midway point oppose each other leading to an increase in the van der Walls pressure in between, pushing the bodies apart, by overcoming the pressure pushing them together.

As mentioned in Lie #22 about inertia, inertia is a form of self-induction. Mechanical motion causes mechanical dipoles to rotate, and the rotation of mechanical dipoles causes objects to move. This is equally true with rotating objects, as it is with objects moving in a straight line, so conservation of angular momentum follows the same basic principle of self-induction.

Rotating objects produce a mechanical-magnetic-like field that leads to pressure changes just like magnets do in electromagnetic theory. The difference is that the velocities of electric currents that produce electromagnetic fields are close to the speed of light and mechanical rotation we normally see is substantially slower. This means that the mechanical-magnetic forces are generally quite small in comparison to electromagnetic forces.

The idea that physics explains mechanical motion is a lie. There is no explanation for how spinning objects such as tops and gyroscopes push against the quantum field. Once we understand that a mechanical dipole must exist, and interacts in accordance with a form of Maxwell's equations and Casimir's theory, it becomes obvious that non-contact mechanical motion is due to pressure differentials in the aether as part of the extended Casimir Force.

[45] A.E. Dolbear, *Modes of motion; or, Mechanical conceptions of physical phenomena.* Boston, Lee and Shepard (1897)

Lie #25: Dark Energy is not a Force

The 25th of the greatest lies in physics is dark energy is not a force. The hypothesis of dark energy was confirmed when astronomers were able to show that the rate of universal expansion is accelerating. Three astrophysicists won a Nobel Prize for their discovery.[46] Since there is no force theory in the standard model to account for accelerating expansion, physicists recognized that there was something unknown going on and gave it the name dark energy. This is a mysterious energy that pulls galaxies apart throughout the deepest regions of space.

According to Newton's second law, **F** = **ma**, a force acting on a mass causes it to accelerate. The opposite is also true; if a mass is accelerating, there is a force acting on it. Dark energy is not merely energy; it is a missing fundamental force. Perhaps it would be better to call it the dark force, and I will from this point forward. Now physicists do discuss forces in energy units quite often, but referring to this force as dark energy is an intentional obfuscation of the truth. Physicists are lying, and pretending that this missing fundamental force is not as big a problem as it truly is.

The unfortunate consequence of this lie is that the missing force problem is not getting the attention it deserves. Most attempts to fix this problem have been patches to existing, unworkable theories. Proponents of new, previously unknown force hypotheses are branded as cranks. Standard model physicists have decided that any force model beyond the standard model is fringe physics, even in this case, when observations tell us that there is a previously unidentified force present.

Fortunately, a force causing matter to be pushed away from other matter is entirely consistent with a mechanical force theory that explains mechanical forces in a similar fashion to electromagnetic forces. In electromagnetic theory, like charges repel, so in a similar mechanical force, like bodies of matter also repel. In this way it is possible that the dark force is simply the mechanical force. Leave it to physicists to overlook the obvious.

That means that up to this point there are six fundamental forces. They are the:

1. mechanical force,
2. electromagnetic force,
3. weak force,
4. strong force,
5. gravitational force,
6. dark force.

Saying that dark energy is not a force is a lie. Physicists will never be able to solve the dark energy mystery until they treat it as a new fundamental force. They must be willing to open up the standard model to add a new force, whatever it may ultimately turn out to be.

[46] The Nobel Prize in Physics 2011.

Lie #26: Gravity is a Fundamental Force

The 26th of the greatest lies in physics is gravity is a fundamental force. Do not be too shocked as the reasoning will be clear in a moment. Also note that this is not to say that the sun and planets in our solar system are not pushed together by some force, they are. It is just that it is not as simple a situation as described by the theories of Newton and Einstein.

The reason why the present gravitational theories are not fundamental is dark energy, or more properly the dark force as discussed in the last chapter. Where there is acceleration, there is a force, **F** = **ma**, and presently there is no force in the standard model causing large electrically neutral bodies of matter to move away from each other.

On one hand we have the well-known gravitational force pushing bodies of matter together. This force is most obvious in our solar system. On the other hand, we now know there is a force that pushes bodies apart at an accelerating rate.

There is, therefore, no way of avoiding the conclusion that Newtonian gravity and Einstein's theory of general relativity are the summation of at least two forces. A force comprised of two distinct and separate forces is not a fundamental force. It is the two forces, whatever they are, which are more fundamental.

One of those forces can be thought of as high-energy gravity and the other the dark force. These two forces act in opposing directions and are nearly equal in magnitude. Note that this also explains one of the long unanswered questions in physics as to how Newtonian gravity is such a weak force. It is actually the difference between two forces that are much stronger by themselves. They are possibly closer in magnitude to electromagnetic forces, as we might expect if all forces could be unified into a single theory.

It is sad that such an obvious problem with gravitational theory has been overlooked. It is not as if the accelerating expansion of the universe is unknown to most physicists. Any physicist with a capacity for critical thinking should have instantly been aware of the consequences of the existence of this previously unrecognized accelerating force.

Fortunately, we can still use Newtonian gravity for calculation purposes while we attempt to understand the true nature of this two-component force. At least on the scale of the solar system, the engineering level math does not need to change. In terms of the physics, we may be able to describe the high-energy gravity force in a way that is similar to Newtonian gravity but with a different constant. General relativity is more problematic and will be the subject of later chapters. A curved space model does not work when we have to sum two forces.

The question of the nature of the dark force is also open. Certainly, if the mechanical force is a field force, as it appears, then that would explain the dark force as well. Scientists have considered other forms of dark energy, which brings up the idea that gravity may be a combination of three or more forces. As scientists, we have to be open to all possibilities, while at the same time seeking the simplest solutions.

There is still a small group of physicists who think the acceleration measurement is an error. That may not get us off the hook though, as a mechanical field force implies that there is a force causing matter to move away from matter. The existence of the mechanical Lorentz force in spiral galaxies tells us we must include a long-range repulsive force in our theories.

There is the additional question as to how high-energy gravity is stronger over shorter distances—on the scale of solar systems and galaxies—and the dark force is stronger over intergalactic distances. That will be left as an open question for the moment as it has to do with the way those two forces are transmitted through the zero-point field.

So now we must amend our list of the fundamental forces now without gravity. They are the:

1. mechanical force,
2. electromagnetic force,
3. weak force,
4. strong force,
5. dark force,
6. high-energy gravity force.

The statement that gravity is a fundamental force is a lie. The existence of the dark force tells us that gravity must be a force composed of two or more forces.

Lie #27: Dark Energy Does Not Cause Expansion

The 27th of the greatest lies in physics is dark energy does not cause expansion. While you may not find physicists explicitly stating this lie, it is an obvious extension of the dark force problem. In any article about the cosmology of the universe, the problems of expansion and acceleration are treated separately. Their assumption is that somehow the acceleration happens while not changing the mechanism behind the expansion.

As discussed in the last two chapters, if there is acceleration, there is a force. This force may be new to us, but it is not a new force. We are talking about a fundamental force, so it has always been acting on all matter, for all of time. That means that not only does the dark force account for the universal acceleration, it must also be responsible for some or all of the intergalactic expansion.

The idea that dark energy does not cause expansion is a lie. The dark force is responsible for accelerating expansion, and is also responsible for expansion in general, or at least some significant fraction of it. Mainstream physicists are lying to themselves and others whenever they fail to consider the dark force's contribution to intergalactic expansion.

Lie #28: Non-Space

The 28th of the greatest lies in physics is non-space. If you are like most people you have probably never heard of non-space as there is not an official term for it in physics. It is a type of space that is not really space. Why would someone want to invent space that is not space? The answer to that question lies with the hypothetical big bang model.

The inventors of the big bang model assumed the big bang could start at a point with a lot of energy. Then the energy would expand, particles would form, and eventually matter would coalesce into atoms, stars, and galaxies. But wait, that meant that the entire universe was a point to start with and initially very small. What was outside the universe?

The answer they proposed was nothing, and it was not only an empty space sort of nothing, it was not even space. As general relativity theory was being developed around that time, physicists were aware of the concept of space-time, but here was empty space outside the universe, so their initial thought was that it had neither space nor time. It did not have any dimensions at all. Of course, since it was outside the universe it was not detectable or measurable either.

You can think of it like you think of blowing up a balloon, where the balloon is the entire universe, and outside it is pure nothingness. As the balloon is blown up and gets larger, the size of the universe increases, but no matter how big the balloon gets, it is always surrounded by nothingness.

What they had invented was a nothingness concept we can call non-space. And of course, this non-space cannot have quantum fluctuations in it either. Quantum fluctuations have wavelengths and frequencies; they have spatial dimensions and time. So, non-space must be quantum fluctuation and aether free. And the aether deniers rejoiced.

But wait a minute; are not these people supposed to be physicists? Physicists are supposed to make physical and mathematical descriptions of things that are physically real. Non-space has no dimensions or time so it is not measurable, and is not physical. Non-space is outside of our universe, so it is not observable, not detectable as well as not measurable. It is not even wrong.

Indeed, non-space has never been observed in our universe. All the space in our universe is real space. Not only that, all the space in our universe contains quantum fluctuations, so it has spatial dimensions and time. All space in our universe has aether.

To top it off non-space is not even necessary as we can come up with a comprehensive theory without it. The non-space hypothesis is an unnecessary complication. It is more like a religion than science. Non-space is a non-scientific hypothesis is since:

    A. There is no physical evidence for it.
    B. It is unnecessary to describe the universe.
    C. It adds unnecessary new complications to the theories of our universe.
    D. It can never be detected.

Non-space is a lie. There is a term that applies to something that is not observable, not detectable, and not measurable. It is called fiction. Or, in this case, since it was scientists inventing this, it is science fiction. There is no such thing as non-space.

Lie #29: The Big Bang Can Ignore Dark Energy

The 29th of the greatest lies in physics is the big bang can ignore dark energy. It has been many years since it was determined that the rate of expansion of the universe is accelerating and the Nobel Prize for that work was handed out years ago. In the mean time the big bang theorists have failed to incorporate the unknown dark force into the big bang model. Or, perhaps no one was willing to publish those results, since they are not favorable to the model.

In the best of situations, the hypothetical big bang model is very fragile, as it has been delicately balanced. One of the fine-tuning problems with the big bang is known as the flatness problem. Using gravitational theory, the density of the universe is precisely balanced between slow steady expansion and collapse. Within the scope of the big bang model, if there were slightly more mass and energy, the universe would have collapsed due to gravity, and if there were slightly less mass and energy, it would have expanded at a much greater rate. To achieve this delicate balance, the initial mass and energy had to be exact to many tens of orders of magnitude. This is in essence statistically, and consequently scientifically impossible, and yet physicists insist on supporting the big bang hypothesis.

Now we have to add the dark force to this delicately balanced model, but if you attempt to add even a small new force into the big bang equations, they no longer work. Every physicist, assuming they have even a basic understanding of force theory, should have recognized this instantly. But instead, they simply do not bring it up in the hopes that the problem will somehow go away, pretending that the dark force somehow does not have to integrated into the big bang model.

Someone could estimate the strength of the dark force based on the acceleration rate. Then they could incorporate this hypothetical dark force into the big bang calculations. Keep in mind that the dark force is not something new, it is something that has acted on all bodies for all time. So if you believe the universe is 13.7 billion years old, the dark force has been accelerating galaxies and other matter away from each other for 13.7 billion years.

That begs the question of how much of the intergalactic velocity, the universal expansion, is be due to the dark force? Is the answer 50%; is it 100%, or perhaps some percentage in between? I will go out on a very short limb and predict it is 100%.

In big bang theory the intergalactic expansion is said to be due to the big bang. Under that theory the initial velocities are thought to be very large and have slowed over time. How much of the expansion due to the hypothetical big bang event would be left if you added the dark force to the calculation? The answer is probably 0%.

The idea that the big bang can ignore dark energy is a lie. The missing fundamental dark force responsible for the accelerating expansion almost certainly accounts for the present rate of expansion of the universe. There could not have been much of a bang to the hypothetical big bang.

Lie #30: The Big Bang Can Exceed the Speed of Light

The 30th of the greatest lies in physics is the big bang can exceed the speed of light. The speed of light limit for bodies of matter is one of the basic experimentally confirmed principles of physics. It is impossible for any body of matter to move at the speed of light. And yet, physicists pretend that it is somehow OK for the big bang model to violate the speed of light limit.

In the simplest version of the hypothetical big bang model, the universe starts as a singularity, a single point in space, or at least something close to being an infinitely dense packet of energy. Then after the big bang happens, the newly produced matter expands outward.

But wait a second. Once all that energy, with or without matter, pops up out of nothing, it instantly forms a black hole. And, as with any other black hole it is impossible for matter or energy to escape its physical boundaries. The matter and energy inside the black hole cannot move. Matter and energy would have to exceed the speed of light to expand at all.

This is an even greater problem than just a single black hole, since super dense matter and energy, like a neutron star, is thought to become a black hole at approximately 3 times the mass of our sun. That means the big bang has to be split into very small chunks of matter and energy, well below the black hole threshold.

All the standard models for the beginning of the big bang violate the speed of light limit with respect to black hole formation. **So, a lie within the lie is that physicists think that all the energy of the universe would not produce a black hole.**

As an aside, if the mass estimate for the visible universe of 10^{56} grams is correct, it would form a black hole with an event horizon radius of 15.7 billion light years.[47] Keep in mind that calculation is based on what could be a low mass estimate, so it may be much bigger. Either gravity does not follow the principles of gravitation on a universal scale, or the standard model for the inside of the black hole is terribly wrong, or both. If you think both, go to the head of the class.

Assuming we could violate this basic principle of physics, and get matter and energy outside a black hole, we still have a problem. The rate of expansion may approach, but must be less than, the speed of light. Then as stars and galaxies form, their light is transmitted though the universe at the speed of light. And, to be clear, light does not move faster than the speed of light either.

If we consider a hypothetical, 13.7 billion-year-old universe expanding from a point at nearly the speed of light, after 6.85 billion years there will be galaxies at the edge of the universe emitting light in all directions. Now say we have an observer still near the original point of origin, someone like us, but not us, since we could not be so lucky as to be exactly in the center. It would take an additional 6.85 billion years for the light from the edge of the universe to get back to the point of origin. So, when that observer sees a galaxy that is 6.85 billion light years away, the hypothetical big bang universe is already 13.7 billion years old.

Of course, that galaxy, if it continued to move at nearly the speed of light would now be 13.7 billion light years away from the point of origin, but the observer at the point of origin could not see that light for another 13.7 billion years, when the universe is 27.4 billion years old.

Astronomers have observed nearly countless objects that are more than 6.85 billion light years away based on the Hubble scale, which means the hypothetical big bang model starting at a point singularity violates the speed of light limit. Not only does it violate the speed of light limit for matter, in order for objects near the edge of the universe to be visible to us now, the speed of light limit for light also has to be violated.

Note the caveat of a big bang starting at a singularity. Some physicists say it does not have to be a singularity. In order for light to travel 13.7 billion years from 13.7 billion light years away, a hypothetical large big bang would have had to instantaneously fill a sphere with a radius of at least 13.7 billion light years. It really needs to be larger, since it is unlikely that we are in the middle, although it does appear that way. It probably appears that way for some reason other than us being exactly in the center.

If we consider such an alternative big bang model that instantaneously fills the volume of the known visible universe, that still does not explain expansion, at least not without going back to the dark force. The other question is, if it fills all space simultaneously; why not simply consider a steady state universe model? Matter is produced somehow. Why not work on that problem first and see where it leads? Until we know how matter is produced without antimatter we are just guessing about the cosmology of the universe.

This problem gets even worse when we consider that astrophysicists now claim that the universe is 93 billion light years in diameter, or perhaps even larger.[48] There is no way for the universe to be that size unless it started that size or larger.

The idea that the big bang model can exceed the speed of light limit is a lie. It is one of the greatest lies in physics. The big bang model requires that the speed of light limits for matter and light be violated.

---

[47] R. Fleming, "The Universe's Large Black Hole Problem." GSJournal.net. 23 Jan 2019.
[48] I. Bars, J. Terning, *Extra Dimensions in Space and Time*, Springer pp27, 2009.

Lie #31: The Inflation Hypothesis

The 31st of the greatest lies in physics is the inflation hypothesis. As pointed out in Lie #30 the big bang model requires that the speed of light limit be violated. We see, according to the Hubble scale, galaxies that are more than 6.85 billion light years away from us. For that to be true with the universe starting from a point, the matter would have to have moved faster than the speed of light. And, for objects to be visible at 13.7 billion light years away, light would have to move faster than the speed of light. Of course, both of these light speed violations are impossible.

In perhaps the single greatest act of self-deception known to physics, the inflation hypothesis was proposed to make the impossible possible.[49] The speed of light limit can now be safely violated according to proponents of the inflation hypothesis.

In the inflation hypothesis, the space of the universe is said to have expanded in an instant, and the matter just went along for the ride, never moving faster than the speed of light relative to space. So not only was there non-space outside the universe there was also a quasi-non-space inside the universe as well. And this quasi-non-space could change physical dimensions and expand rapidly. This way the more delusional physicists could con themselves into thinking that the speed of light limit was not violated.

These physicists are of course quantum fluctuation deniers, or they could never believe such nonsense. Real space contains quantum fluctuations, so in order for any physically real space to expand at a rate much higher than the speed of light, there has to be enough energy to push all the quantum fluctuations at that speed. This superluminal matter also has to somehow overcome the normal resistance to motion due to quantum fluctuations.

And yes, inertia is still a form of resistance to motion and the speed of light limit is the same for electrically neutral bodies as electrically charged bodies and light. The quantum field's resistance that gives us permittivity and permeability, and the speed of light limit is the same for all bodies.

Keep in mind that a cubic centimeter of vacuum has at least the energy equivalent of 10^{95} grams of matter which is much greater than the mass of the entire visible universe under the big bang model. There is not enough energy in the entire hypothetical big bang to expand even a cubic centimeter of quantum fluctuations, even less inflate it at impossible speeds.

There is no such thing as a quasi-non-space, whatever you want to call it, within our universe. Such a thing has never been observed. Everywhere we look there is space, there is the zero-point field, and there are quantum fluctuations. The inflation hypothesis is not physics. It is science fiction, and it is a lie.

[49] A. Guth, The Inflationary Universe: The Quest for a New Theory of Cosmic Origins, Perseus, 1997.

Lie #32: The Cosmic Microwave Background Proves the Big Bang

> ...in the case of radiation in thermal equilibrium with charged particles we may return to the considerations of Planck, Einstein, and Bose, which provide us with Planck's formula for the intensity of radiation (in equilibrium with matter) for each frequency v. ...we shall be seeing the huge relevance that this formula has to cosmology as it is in extremely good agreement with the radiation spectrum of the cosmic microwave background.[50]
>
> Roger Penrose, 2016

The 32nd of the greatest lies in physics is the Cosmic Microwave Background (CMB) proves the big bang. Here is another ridiculous assertion that could only be dreamed of by quantum fluctuation deniers. For background, two American radio astronomers Arno Penzias and Robert Woodrow Wilson accidentally discovered CMB in 1964 when they noticed background noise in their horn antennae that was uniform across the sky.[51] The microwave wavelength they measured was roughly equivalent to a temperature of 3 Kelvin (K), currently accepted to be 2.7K. Penzias and Wilson later were honored with the Nobel Prize for this work.

The idea that the vacuum of space had a non-zero temperature was nothing new. In 1896 Charles-Edouard Guillaume calculated that space had a temperature of 5.6K due to starlight.[52] Sir Arthur Eddington made a similar calculation and derived a temperature of 3.18K, which is very close to the measured temperature.[53]

Later, after the invention of the big bang model, several physicists calculated the residual heat energy from a hypothetical big bang and generally found that it would leave a temperature signature about 10 times greater than the currently accepted temperature. Nonetheless, once the CMB was discovered, the big bang theorists instantly claimed that CMB was proof of the big bang model. They then tweaked their estimates to fit the data. This was the fallacy of wishful thinking in action.

The measurement of microwave emission throughout space actually tells us very little about what causes it without additional information. As mentioned before, scientists were able to detect a Doppler shift indicating that our solar system is moving relative to the CMB, meaning that the universe truly does have a rest frame. Since it is not physically possible for there to be two rest frames, one for aether and one for the CMB, this must be the aether rest frame.

Also note that the only two things we know to be relatively uniform throughout the visible universe are aether and the CMB. The radiation of the CMB must be emanating from the aether in some way as all black body radiation in a vacuum is due to interactions with quantum fluctuations. As Roger Penrose states, the shape of the black body radiation from the CMB distribution matches Planck's prediction. The CMB emits black body radiation through a mechanism involving aether. Empty space could not radiate if it were not filled with aether. Evidence of the CMB proves the existence of a Planck type aether in addition to the aether rest frame.

It is important to note that the CMB comes from all time and all space, since we are looking further back in time the further we look away. The CMB is not just coming from a single point in time long ago in the history of the universe. It is coming from all times in the past and is unchanged. This means that there have been no major changes in fundamental constants that might change the black body spectrum of the CMB over time. Whatever causes the CMB is an ongoing process.

Beyond that we can only say that there must be an underlying process that causes the background temperature that leads to the CMB. The source of that temperature has not yet been identified with any degree of certainty.

The problem with the early light models is two-fold; stars are not uniformly spread throughout the universe, and quantum fluctuations do not absorb photons. There are voids as big as a billion light years diameter. The heat and light from the stars are not uniform, so unless heat propagates through the aether at a velocity much faster than the speed of light, the CMB is not due to light. We would also need a physical mechanism describing how photons interact with quantum fluctuations to heat them and then we need a mechanism to release black body spectrum photons. We do not have either in the standard model.

We generally think that aether, which adheres to the Planck and Heisenberg limits, would not have a detectable temperature above zero, as each quantum fluctuation returns to zero energy. But we may need to reexamine that assumption. Temperature in a gas is a function of the kinetic energy of the gas molecules. Perhaps the van der Waals energy of the vacuum gives it an apparent temperature due to van der Waals induced motion. Then when quantum fluctuations occasionally cross-annihilate, this van der Waals temperature could be radiated, emitting a black body spectrum. Such a physical mechanism, if it has the correct effective temperature, would explain the CMB.

Alternatively, we know that matter does exist, so we also know it must be produced in some way without antimatter. And, we do know that the big bang model does not actually explain a process for matter to be produced without antimatter. **Actually, that is another lie hidden in the big bang model, that it explains the production of matter.**

The physical mechanism behind matter production, whatever it is, may be the cause of the excess temperature. The CMB radiation would then be radiation from that matter, due to its temperature. If that is the case, the uniformity of the CMB favors a uniform matter production process, in both space and time, occurring throughout the universe, rather than a one-time big bang type event.

The statement that the CMB proves the big bang is a lie. The CMB is evidence of some thermal process, possibly zero-point energy or matter production, with perhaps an outside chance of a light absorption phenomenon. It is by no stretch of the imagination, proof of the big bang model. The CMB is, however, proof of a Planck type aether and the aether rest frame.

[50] R. Penrose, *Fashion Faith and Fantasy in the New Physics of the Universe*, pg 195, Princeton University Press, 2016
[51] A.A. Penzis, R.W. Wilson, "A Measurement of excess antenna temperature at 4080 Mc/s." Astrophysical Journal, Vol. 142 pp.419-421, 1965.
[52] Guillaume, Charles-Edouard (1896). "La Température de L'Espace (The Temperature of Space)." La Nature. 24.
[53] Eddington, A. *The internal constitution of the stars*. Cambridge, United Kingdom: Cambridge University Press. pp371-372 1926.

Lie #33: The Big Bang Can Violate Conservation of Energy

The 33rd of the greatest lies in physics is the big bang can violate conservation of energy. Anyone who has studied even a little bit of physics understands that conservation of energy is on a very short list of the most important principles of physics. Energy is neither made nor destroyed, energy can only be transformed from one type of energy to another. Perhaps the most famously known transformation is the conversion of mass to energy following the **$E = mc^2$** formula.

There are far more common conversions involving chemical energy, heat energy, mechanical energy, electrical energy, and nuclear energy. Energy such as chemical and nuclear can be stored and later released to make heat. In electrical plants the heat can heat steam, and the expansion of the steam can be converted to mechanical energy, which can then be converted to electrical energy. Quantum field energy can even move objects through the Casimir effect. Energy conversion of one form or another is common and found in most devices we operate daily.

The one thing we cannot do, however, is make energy from nothing. We can only extract energy that is already present, which includes zero-point energy. The energy may be stored in atomic nuclei, or in chemical bonds between atoms and molecules. We can store and extract energy due to electrical potential differences between atoms, or electrical fields between conductive or semi-conductive layers. In every case the energy is already stored, or we have to add it to a system from another source of energy for later retrieval.

The big bang model is said to start with a singularity, or perhaps a larger scale instantaneous event. In the big bang model, all the energy in the universe comes out of nothingness in an instant. This is a special kind of nothingness. It is not normal space containing quantum fluctuations and zero-point energy. Big bang theorists say that before the big bang there was a sort of non-space space, a space without quantum fluctuations, without dimensions and time, and without energy. You may recall non-space from Lie #28.

Big bang theorists believe they can get something from nothing. They believe that energy can magically appear when there was no energy before. This is a conservation of energy violation on a truly universal scale. Many physicists, whose capacity for self-deception appears limitless, simply ignore this problem.

The reason big bang theorists feel they can ignore the conservation of energy violation may go back to most of them being quantum fluctuation deniers too. If there are no quantum fluctuations, there is no zero-point energy, so their idea of space has no energy. This leads them to think that since matter has to come from nothing, whether it is a big bang model or some other model, then it must be OK to get matter and energy from nothing. It is not OK.

If we recall from Lie #19, mass is intrinsic, the masses of the electron, proton, and neutron are equivalent to the amount of zero-point energy they displace. This also means that the existence of ordinary matter does not change the total amount of energy in space. When space contains only quantum fluctuations, it has one amount of energy. If we introduce an electron, space still has the same amount of energy. Matter cannot exist without zero-point energy, and the existence of matter does not change the amount of energy in space.

It is a lie to say that the big bang model can violate the principle of conservation of energy. Physics requires that there are no violations of the principle of conservation of energy, and the big bang model is the largest possible violation of conservation of energy there is. Fortunately, we do know the source of energy behind all the mass-energy in the universe. It is the zero-point energy, and it has been here as long as the universe.

Note that this leads us to a broader version of the lie. **Any cosmological model that does not assume aether filled space existed for infinity, violates the principle of conservation of energy, and is therefore a lie.**

Lie #34: The Big Bang

> *I have never thought that you could obtain the extremely clumpy, heterogeneous universe we have today, strongly affected by plasma processes, from the smooth, homogeneous one of the Big Bang, dominated by gravitation.*[54]
>
> Hannes Alfvén

The 34th of the greatest lies in physics is the big bang. Based on the last few physics lies in this book it should be readily apparent that the big bang is a big lie. An entire book could be, and many have been written about the many lies fabricated to cover up the big bang lie. There are dozens of good reasons why the model does not work. Here are a few.

1. It violates the principle of conservation of energy.
2. The science fiction non-space concept.
3. The black hole problem.
4. The speed of light limit violations.
5. The large size, 10^{93} billion light years, problem.
6. The ridiculous inflation theory.
7. The existence of the dark force.

The Hubble telescope has also given us observational data about highly redshifted galaxies and galactic clusters and superclusters. We see very distant objects that are not young, but are often very old, on the order of 10 billion years or more. And then, based on the Hubble scale, these objects are 10 billion plus light years away from us. Based on these observations alone the universe has to be more than 20 billion years old.

So even if one believes that matter moved faster than the speed of light in the early universe, these objects are still older than the 13.7 billion year age of the hypothetical big bang. The papers that report these observations are odd in the way they challenge the big bang while not challenging it, since the mainstream journals do not allow physicists to openly challenge the big bang model.

There are also voids and walls that may have taken much longer to form. There is a void, based on observations of surrounding stars, that is approximately a billion light years in diameter where few galaxies are visible. If galaxies once occupied that region of space and moved apart at a nominal but fast intergalactic speed of say 1000 kilometers per second, it would have taken hundreds of billions of years for the void to form. In a slow developing, steady state model, the universe must be more than a trillion years old, and that model matches observation much better than the big bang.

There are numerous books, papers and Internet sites that discuss the numerous lies imbedded in the big bang model. Please feel free to research it further. Any scientifically minded person capable of rational thought who studies the big bang model's failings will conclude that the big bang model is a lie.

[54] Quoted in Anthony L. Peratt, "Dean of the Plasma Dissidents." Washington Times, supplement: The World and I (May 1988),196.

Lie #35: Cyclical Universe

The 35th of the greatest lies in physics is the cyclical universe. The cyclical universe model comes about as a consequence of the big bang model. Historically the big bang model walked a tightrope as to whether the universe would expand forever or collapse. A collapsing universe is sometimes referred to as the big crunch.

This is the flatness problem and is the previously discussed fine-tuning problem with the big bang model. If there were slightly more mass, the universe would have collapsed long ago. If there were slightly less mass, the universe would already have expanded at a much faster rate, with no hope of ever collapsing. This balancing act between mass and gravity had to be perfect to a ridiculous number of decimal places. Of course, such a precise result is physically impossible.

In any case, given that one possibility was that the universe would eventually collapse it became obvious that the pattern could hypothetically repeat itself, if you are willing to ignore some basic principles of physics. There could hypothetically be an infinite number of big bangs followed by periods of expansion and collapse to an equal number of big crunches. This is often referred to as the cyclical universe model. The cyclical universe model has an advantage over a single big bang model in that energy from one big bang universe can contribute to the next thus fixing the conservation of energy problem if it is assumed to go on forever.

Once there was scientific evidence supporting the accelerating expansion of the universe, the collapsing universe model was invalidated, along with the cyclical universe model. If the rate of expansion of the universe is accelerating, it is impossible for it to collapse. Even so, many years later references to a collapsing universe or cyclical universe can still be found.

As with so many things big bang related, the lie is much worse than that. If we consider matter coalescing, it will form a black hole. At that point, matter and energy will not be able to cross the event horizon except for the event horizon growing with the black hole's increase in mass. The matter added to the black hole is still forever frozen in time at the location of the event horizon when it fell into the black hole. **The concept of a big crunch is a lie.**

Once this black hole forms, there will also be no mechanism allowing it to become a big bang again, as the matter and energy will be stuck at or inside the event horizon of the black hole forever. This is complicated by the fact that if the gravitational constant G is actually constant, the entire visible universe should already be inside a black hole.

Of course, the big bang model is a lie which makes the cyclical universe model an even worse lie. For the cyclical universe to be true, the universe model would have to be a specific type of big bang model that starts with a singularity, or point in space and time. As discussed previously, such a big bang model violates the speed of light limits for black holes, bodies of matter, and for light. The singularity will be a black hole that cannot expand, in addition to having numerous other problems.

Lie #36: The Speed of Gravity is the Speed of Light

The 36th of the greatest lies in physics is the speed of gravity is the speed of light. Here is another lie brought to us by quantum fluctuation deniers. First, they deny the existence of quantum fluctuations and have to make the speed of light an intrinsic property of light, rather than due to interactions between photons and the zero-point field. From there they decide that the maximum speed of everything is the speed of light by default.

In Newton's day up through the early 20th century it was well known that the speed of transmission of the gravitational force had to be instantaneous or nearly so, or the orbits we see would be unstable. If the speed of gravity were the speed of light, by the time the force signal was received, the bodies would have moved, leading to forces that are not directed at each of the bodies' current positions. This is simply illustrated in the drawing below where a body is at position A when the force is transmitted, but is in position B when it is received so the body is pushed in the wrong direction.

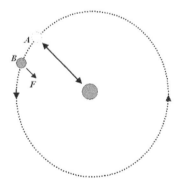

Figure 36-1 A star in the center with a planet in orbit. The arrows illustrate that if the gravitational force is not instantaneous it is not directed toward the other body.

Astronomical observations verify that the gravitational force on the Earth is not directed at the Sun's position based on the light we see, which is of course delayed. The gravitation force is directed at the Sun's actual position.[55]

The consequence of this fatal flaw in the speed of gravity equals the speed of light lie, is planets in orbit would gradual get further and further away. In the case of elliptical orbits, there also would be significant additional precession as the bodies get further apart. There can be no stable orbits in the slow gravity scheme. Tom van Flandern did a good analysis where he calculated that the speed of gravity must be at least 2×10^{10} times the speed of light.[55] It is likely many orders of magnitude faster than that.

Mainstream physicists have come up with several self-deceptive ways of explaining away the problem. In a curved space model, they say the bodies follow a geometrical curvature of space and there is no real force being transmitted. They say it is no different than bodies moving parallel past each other.

Then you ask how space knew which way to curve, they say that all space instantaneously knows where all the matter is from everywhere in the universe. Did you catch that? Instantaneous in this case means that information has to be transmitted much faster than the speed of light, so their theory equates to the speed of gravity being much faster than the speed of light anyway. They have rephrased their lies a multitude of ways to deceive themselves and others, but if you look close enough, you find that they are trying to hide the part where the force, or information about the force, was transmitted faster than light.

Physicists who believe this nonsense will also claim to have experimental evidence to overcome the classic evidence that the speed of gravity is instantaneous. If one studies these so-called proofs you learn the experiments are actually measuring the speed of light. And they are typically not measuring gravity at all, but rather a component of the mechanical force.

Then we have to consider that Newtonian gravity and general relativity must be the sum of two opposing forces, the force pushing bodies together—high-energy gravity—and the force pushing bodies apart responsible for accelerating expansion of the universe—the dark force. Both high-energy gravity and the dark force have to be transmitted at velocities much faster than the speed of light for stable orbits to occur. If both forces propagated at the speed of light there really would be no such thing as a stable orbit.

The claim that the speed of gravity is the speed of light is a lie. It is another lie brought to us by the aether deniers.

[55] T. van Flandern, "The Speed of Gravity--What the Experiments Say." Meta Research, Physics Letters A, 250:1-11, 1998.

Lie #37: The Speed of Electromagnetic Fields is Infinite

The 37th of the greatest lies in physics is the speed of electromagnetic fields is infinite. The mainstream physicists in the crowd are now thinking "we do not say that." And, it is quite true that one will never find that statement in any textbook. If anything is said about the speed of propagation of electromagnetic fields it is claimed to be the speed of light.

The idea that the speed of electromagnetic fields is infinite is implicit in physicist's assumptions and equations, even if it is unstated. As with so many of their other great lies, this one is once again due to the quantum fluctuation deniers. Since they refused to accept the evidence for zero-point energy which is the medium for photon transmission, they had to make electric and magnetic fields intrinsic properties of photons, along with the speed of light.

The word intrinsic should have warned them that they were making a big mistake, another lie within lies. On one hand physicists insist that electromagnetic fields propagate at the speed of light and on the other hand they insist the propagation speed is infinite, just there intrinsically. This is similar to the situation with gravitational force transmission. As with Newton's equation for gravity, Maxwell's equations for electricity and magnetism assume that force transmission is instantaneous, or nearly so.

To illustrate, consider that when a photon is produced it has electric and magnetic fields that are perpendicular to each other. These fields are part of the photon's wavefront. Since the photon originates from a new point each half-wavelength, the electromagnetic fields have to propagate at an infinite speed in order for them to be detectable. If not, they would not be detectable beyond a half-wavelength.

If the electric and magnetic fields of a photon propagated outward from the central dipole at only the speed of light, the fields would not be detectable outside each individual dipole's light cone. The electromagnetic field does not know in advance of when the photon will be absorbed.

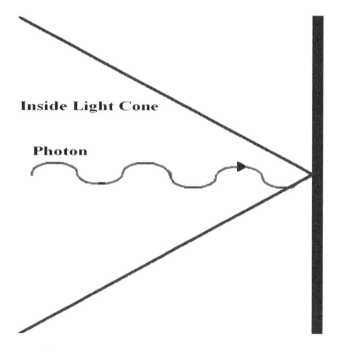

Figure 37-1 A photon is absorbed at a surface. If there is a speed of light limit for electromagnetic field propagation, only the field energy within the light cone and adjacent to the last dipole could be absorbed.

When a photon is absorbed, say after traveling a billion light years, only the central dipole energy would be absorbed. This energy would only come from a region of space within a half-wavelength of the photon. Under the speed of light limit, the rest of the electromagnetic field would be lost and have to be absorbed into the aether or nearby matter.

How do the fields know they must cease to exist? Based on the hypothetical speed of light limit for propagation of the fields, communication between the photon dipole and the extended fields would be impossible. In order for the electromagnetic fields to collapse at the proper moment, gathering all of the photon's energy, the speed of propagation of those fields has to be infinite. That is the implicit assumption one makes once one buys into the standard model view of the photon.

Since quantum fluctuations do exist, those dipoles form the electric and magnetic fields through their polarization and rotation. These fields propagate dipole to dipole through space. It is physically impossible for this propagation to occur at infinite speed, even though it is extremely fast.

That is why the implicit assumption of infinite speed of propagation of electromagnetic fields is a lie. It is part of the lie that electric and magnetic fields are intrinsic properties of photons.

Lie #38: Electromagnetic Fields Propagate at the Speed of Light

The 38th of the greatest lies in physics is electromagnetic fields propagate at the speed of light. After the last chapter this lie is pretty obvious. As usual it is the quantum fluctuation deniers who are responsible for this lie. And, once again this lie started with the false assumptions that the electric and magnetic fields and the speed of light are intrinsic to photons, rather than an interaction with quantum fluctuations, the photon transmission medium.

The assumption that electric and magnetic fields move at the speed of light is easily disproved. In the first case we have the same problem we have with gravity as explained in lie #36. Maxwell's equations, like Newton's gravity, actually contain an implicit assumption that forces propagate instantaneously, and Maxwell's equations make excellent predictions about the curved paths that charged particles follow in electric and magnetic fields. If forces only traveled at the speed of light, those paths would be different. Once the force information is received, the object transmitting that information would be in a different location causing the forces to act in a different direction.

If that alone is not convincing, a simple analysis about how fields physically propagate through the aether should help. As we know due to the Casimir effect, quantum fluctuation dipoles exist. As dipoles, they can be polarized and rotate in response to moving charges, and produce van der Waals forces and torques. When quantum fluctuation dipoles polarize, they form an electric field and when they rotate, they form a magnetic field. Both of these fields propagate dipole to dipole through space.

On a mass equivalent basis, there are ~10^{95} grams of zero-point energy per cubic centimeter, which is ~10^{94}

times greater than the typical matter on Earth. That means there is a similar ratio of charges, such that from our perspective, the vacuum has a nearly infinite number of charges, and a nearly infinite capacity to form electric and magnetic fields. Each dipole can rotate 180 degrees at the speed of light, but only a small fraction of the quantum fluctuation dipoles actually have to, in order to form electric or magnetic fields.

More importantly, quantum fluctuations do not have to rotate a full 180 degrees for an electric or magnetic field to exist and propagate; they only have to rotate a small fraction of a degree. If they had to rotate 180 degrees to produce a field, fields would propagate at the speed of light, but since fields propagate without that much rotation, they propagate much faster than the speed of light. Even rotating $1/10^{10th}$ of a degree produces an electric or magnetic field far stronger than anything we can make. Fields can and do propagate at greater than 10^{10} times the speed of light through the aether.

How fast is the speed of propagation? That is hard to tell. Given the large disparity between the energy of matter and the energy of the zero-point field, electromagnetic fields may propagate at 10^{94} times the speed of light or faster. Note that while we cannot currently measure such a small difference from an infinite velocity, it is still not infinite.

The assumption that the speed of propagation of electric and magnetic fields is the speed of light is a lie. The dipoles that fill the aether allow those fields to propagate at a rate much faster than the speed of light.

Lie #39: The Horizon Problem

The 39th of the greatest lies in physics is the horizon problem. The horizon problem is sometimes referred to as the homogeneity problem. It comes about because the CMB and other phenomena are uniform throughout the visible universe; when because of the hypothetical speed of light limit, parts of the universe have not had time to be in contact with each other in a big bang type model. Under this hypothetical speed limit this problem would occur with any cosmological model that assumes a universe that is not infinitely old.

In lie #32 it was mentioned that the thermal uniformity of the aether, as demonstrated by the measured CMB radiation, indicates that heat, from whatever cause, causes the CMB. The last chapter tells us that electromagnetic fields propagate much faster than the speed of light, so that means that van der Waals forces also propagate much faster than the speed of light. The jitter of quantum dipoles in response to van der Waals forces is the closest thing to classical heat that is generated by aether.

This tells us that heat can be transferred at a rate much faster than the speed of light. Van der Waals forces do not depend on quantum dipoles rotating 180 degrees, as it does with photons. The thermal properties of the CMB, as with forces, can propagate through aether with a speed much faster than that of light. The uniformity of the CMB and other properties of the universe is not surprising, regardless of what causes the CMB.

The horizon problem is a lie. But even so, the universe can still be infinitely old so that energy is conserved.

Lie #40: General Relativity Explains Gravitational Acceleration

The 40th of the greatest lies in physics is general relativity explains gravitational acceleration. The general relativity model evolved due to rejection of quantum fluctuations and failure to discover there is a fundamental field force underlying basic mechanics. Due to these incorrect assumptions, general relativity has numerous flaws, with one of the biggest being that it does not actually explain the mechanism responsible for gravitational acceleration.

To expose this oversight all we have to do is consider a two-body problem where two bodies are in space adjacent to one another, but not initially moving relative to each other. General relativity describes that there is a path between them that they will follow. Without any relative tangential velocity, or other bodies nearby, this path is a straight line between the two bodies rather than an orbit of some kind.

General relativity, however, does not show us how the bodies are pushed so that they accelerate toward each other. As such, in general relativity the two bodies would stay where they were indefinitely. Even if there are other bodies nearby, general relativity does not say how those additional bodies accelerate or push on the first two. If a hypothesis cannot even explain a simple two-body problem then it does not deserve to be called a theory. Gravitational theorists need to stop calculating and find the physical explanation.

Now it is important to note that Newtonian gravity does not tell us where the acceleration comes from either, so it is not like we can revert to an older model. Newton's equations are sufficient for determining the results of acceleration within the solar system, but not how the acceleration occurs.

This problem is compounded considerably when we note that there also appears to be a force pushing bodies apart in addition to the one pushing them together. Newtonian gravity or general relativity would need to explain how both these forces push on bodies for those models to be successful. Those two forces must be integrated into a single model.

When we look for where the push comes from, the only thing we have in space is the quantum field and the only significant thing in the quantum field is quantum fluctuations. As such, the push has to come from pressure due to interactions between quantum fluctuations. This leads us back to the Casimir effect. Acceleration in one direction or the other is due to quantum pressure differentials.

It turns out that Casimir was not even the first to suggest this. The first was a young Swiss scientist named Nicolas Fatio de Duillier, a contemporary of Newton's.[56] He originated the push theory of gravity, also called the shadow theory of gravity. This theory was later popularized by Georges-Louis Le Sage, and often is referred to as Le Sage's theory of gravity. Le Sage was well aware of Fatio's work, but did not give him credit.

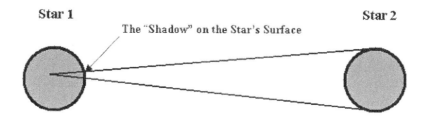

Figure 40-1 An illustration of the shadow from the shadow gravity part of the theory.

Fatio's idea was that space is filled with corpuscles, and as these corpuscles move through space, they randomly strike each other and astronomical bodies, pushing them. If we consider two bodies, the first body shadows the second body from the corpuscular pressure force coming from behind the first body, and in this way, there is more force pushing the bodies together than pushing them apart. The concept is illustrated in the two figures.

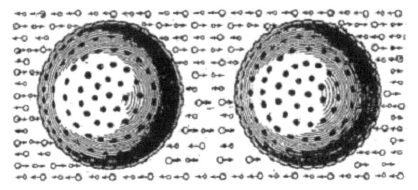

Figure 40-2 An illustration of the Fatio's Push Gravity Theory in a drawing by Le Sage.

Like aether theory in the Michelson-Morley age, Fatio's theory suffered due to kinetic interpretations of the corpuscles' behavior. Lord Kelvin and others pointed out that a purely kinetic interaction between the corpuscles and astronomical bodies would generate heat.[57,58,59,60] It was not until Planck and Casimir that we had a clearer definition of what those corpuscles were. The corpuscles are quantum fluctuations and they are dipoles. As with the Casimir force, quantum fluctuations do not cause bodies to heat.

Since gravity, as we have known it, is actually a two-component force, the Fatio-Casimir model applies to the high-energy gravity part. Van der Waals forces between quantum fluctuations produce pressure that pushes on all bodies. Due to the shadowing effect, the pressure between two bodies pushing them apart is reduced relative to the pressure pushing them together. The net effect is that bodies are pushed together. The mechanical force explains the dark force as previously described in Chapter 24.

It is also important to note that Fatio's theory is a true quantum gravity theory, as it is based entirely on quantum electrodynamic principles. It also unifies the gravitational force with the electromagnetic and mechanical forces. The existence of the Casimir force tells us that the Fatio-Casimir force must exist; it is not even a matter of debate.

Instead of trying to figure out how bodies accelerate due to gravity, mainstream physicists have pretended that their theories explain the acceleration when that is a lie.

There is one last question that has so far prevented the Fatio-Casimir push gravity theory from being accepted and that is how is the *$1/r^2$* varying van der Waals force generated? That is the topic of the next lie.

[56] N Fatio de Dullier, "De le Cause de la Pesanteur" (ca 1690), Edited version published by K. Bopp, Drei Untersuchungen zur Geschichte der Mathematik, Walter de Gruyter & Co. pg 19-26 1929.
[57] W. Thomson, (Lord Kelvin), "On the ultramundane corpuscles of Le Sage", Phil. Mag. 45: 321–332, 1873.
[58] J.C. Maxwell, "Atom", in none, Encyclopedia Britannica, 3 (9th ed.), pp. 38–47(1875).
[59] H. Poincaré, "The Theory of Lesage", The foundations of science (Science and Method, 1908), New York: Science Press, pp. 517–522, 1913.
[60] R.P. Feynman, The Character of Physical Law, The 1964 Messenger Lectures, pp. 37-39, 1967.

Lie #41: Van der Waals Forces do not Explain Gravity

The 41st of the greatest lies in physics is van der Waals forces do not explain gravity. As discussed previously, van der Waals forces are due to interactions between dipoles. In this case the dipoles are quantum fluctuations. The difficulty is that van der Waals forces between electric charge dipoles have a short range. Short-range van der Waals forces do not have a measurable effect at the distances between stars, as they decrease at a rate much faster than $1/r^2$. In order to explain gravity, some of these forces must follow the inverse square law.

Van der Waals forces are, however, the only known type of force between quantum dipoles that can generate the push needed to cause gravity. Consequently, at least one scientist has attempted to derive a van der Waals force that follows the inverse square law.[61] That attempt was not successful.[62] Consequently the Fatio-Casimir theory of gravitational action has not achieved broad acceptance.

Part of the problem is physicists' general beliefs that forces work by magical action at a distance, and magical jetpacks, or whatever, that produce propulsion. If we attempt to understand how the vacuum pushes objects like magnets, we must recognize that some kind of extended Casimir force truly does exist.

There is quantum fluctuation pressure pushing on all bodies and that pressure is responsible for all motion due to electromagnetic, mechanical, and gravitational forces. Each of these forces in turn varies with distance following the inverse square law. That means that there must be a fundamental van der Waals pressure throughout space that is consistent with the inverse square law.

Consequently, we are not dealing with a question of if there is a $1/r^2$ varying van der Waals pressure in space. The physical evidence for the inverse square law as it applies to fundamental forces tells us this $1/r^2$ varying pressure exists. The physical evidence for it is incontestable, as we have no other explanation for motion consistent with the existence of quantum dipoles.

The question is then only a matter of how we derive it mathematically. The previous failures to derive the force simply tell us that we have not yet discovered the correct physical explanation for the force within the scope of electromagnetic van der Waals force theory.

The reason for this difficulty is that physicists, in general, failed to recognize that mechanical forces behave as a field force and therefore there must be a second dipole, an electrically neutral dipole. This dipole must be different from the positive and negative electric dipole. Mechanical objects, such as a spinning top, do not have electric charge and do not produce electromagnetic fields.

As mentioned in other chapters, the most likely mechanical dipole is Dirac's positive and negative 'energy' dipole, matter and antimatter. Deriving a $1/r^2$ varying van der Waals force is simple once we recognize that there are two separate types of dipoles and that each particle has two types of charge. It does not even matter if the mechanical dipole is considered to be something other than matter and antimatter; the existence of a non-electric dipole of some type is sufficient to solve the problem.

The reason for this is simple, as there are two types of dipoles with two different charge polarity combinations. One dipole has positive electric and negative mechanical (antimatter) charge on one end and negative electric and positive mechanical (matter) on the other end. This dipole has charges like an electron-positron pair.

There is also a second type of dipole that has positive electric and mechanical (matter) charge on one end and negative electric and mechanical (antimatter) charge on the other end. This dipole has charges like a proton-antiproton pair.

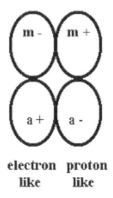

electron proton
like   like

Figure 41-1 Two dipoles, one electron-like and one proton-like, adjacent to each other with their electric dipoles canceling and their matter dipoles repelling

The importance of this is that these two different dipole types always repel each other no matter what their orientation. Remember that electrons are repelled from protons when they get close. If they line up such that their positive and negative electric charges are nearby, their mechanical charge is in a repulsive alignment, (matter to matter or antimatter to antimatter). This is illustrated in figure 41-1. If on the other hand their mechanical charges are aligned with opposites nearby (matter to antimatter) then their electric charges repel (positive to positive and negative to negative).

If one rejects the idea that non-electric mechanical forces are a field force, the short-range repulsive forces between electrons, between protons, and between protons and electrons is enough to generate the $1/r^2$ varying van der Waals pressure force when combined with electrostatic repulsion. Even if one wants to attribute these forces to the Pauli exclusion principle, the proven existence of these repulsive forces between particles of matter is sufficient to explain the $1/r^2$ varying van der Waals force.

This situation leads to an entirely new type of van der Waals force, a force where there is constant repulsion between dipoles no matter what their orientation. It is this new van der Waals force that is the source of the $1/r^2$ varying pressure responsible for gravity, as well as being the source of the inverse square law of electromagnetic and mechanical forces.

The idea that van der Waals forces do not explain gravity is a lie. The electromagnetic force follows the inverse square law, and electromagnetic motion must be due to the extended Casimir force. This tells us that there must be an underlying $1/r^2$ varying van der Waals pressure coming from the quantum field. This van der Waals force is due to the repulsive forces between matter particles as well as those between antimatter particles.

[61] H.E. Puthoff, "Gravity as a Zero-Point-Fluctuation Force," Phys Rev A 39:2333, 1989.
[62] S. Carlip, Phys. Rev., 47: 3452, 1993.

Lie #42: General Relativity Can Ignore Zero-Point Energy

The 42nd of the greatest lies in physics is general relativity can ignore zero-point energy. This lie can be laid at the feet of our greatest quantum fluctuation denier, Albert Einstein. He and others who developed general relativity simply ignored zero-point energy, knowing full well that if they had to add zero-point energy to the theory, it would not work.

Aether has energy equivalent to ~10^{95} grams per cubic centimeter, assuming a cutoff at the Planck length. A cubic centimeter of vacuum contains more energy than the entire visible universe. In the general relativity model, all energy is responsible for the gravitational force. So, if the quantum field energy were included in the general relativity calculation, all the matter in the universe would be compacted by gravity to the highest density form of matter possible, a big black hole. Knowing this, mainstream physicists who favor general relativity theory, which is most of them, have continued to ignore zero-point energy and perpetuate the lie.

Note that under black hole theory one cubic centimeter of vacuum would create a black hole 10^{68} meters in diameter, which is about 10^{42} times larger than the visible universe. So obviously, zero-point energy does not have a gravitational effect on light or matter.

The physicists' lie is further compounded when we consider that, as Dirac first proposed, mass is due to the displacement of vacuum energy, such that mass-energy is equal to displaced quantum field energy. That means that the total energy in space is always equal to the initial zero-point energy in empty space. Consequently, there is no energy gradient in the universe. The concept of relating gravity to the total energy gradient in space is preposterous.

As discussed in the last lie, there is an easy way around the zero-point energy problem as it relates to gravity. Quantum fluctuations exert pressure uniformly in all directions such that a body in otherwise empty space is pushed on equally from all sides. The body is not heated either as proved in Casimir force experiments. Pressure differentials due to Fatio's shadowing effect cause gravitational acceleration.

Baseline quantum field van der Waals pressure is positive and uniform from all directions. In other words, aether gravitates without producing a detectable gravitational force since it pushes equally from all sides. We only detect it when there is a pressure differential.

It is a lie that general relativity can ignore zero-point energy. Quantum fluctuations and aether are real, and they must be accounted for in any gravitational theory.

Lie #43: Curved Space

The 43rd of the greatest lies in physics is curved space. As with so many of the other great lies in physics, this one is also the result of the rejection of aether. Once one considers zero-point energy, it is easy to recognize there is not enough mass-energy or other forms of energy to displace or otherwise curve the vacuum. There is far too much zero-point energy for mass-energy to have much of an effect. This lie is also the result of physicists' failure to recognize that there is a far simpler explanation for the mechanical force phenomena they were trying to explain, such as the precession of the perihelion of Mercury.

The quantum field has mass-energy equivalent of $\sim 10^{95}$ grams per cubic centimeter, assuming a cutoff at the Planck length. Zero-point energy is more than 10^{94} times the typical non-vacuum energy on Earth. The total mass of the visible universe has been estimated to be 10^{56} grams.[63] So how is such a small amount of mass supposed to modify the geometry of the vacuum and its large associated energy? It can't. Space by itself does not even have physical dimensions since space is non-physical. The physical dimensions come from the wavelengths of the quantum fluctuations which are fixed in the aether rest frame.

This lie is further compounded when we consider that the mass-energies of the electron, proton, and neutron are equal to the zero-point energy they displace. This leads us to conclude that the total energy in space is constant. There is no energy gradient to space and thus, the idea of space curvature is nonsense.

From a historical point of view the curved space model was developed because physicists failed to recognize that moving matter produces the same types of forces as moving electric charges (See Lie #24). Rotating bodies such as tops and gyroscopes produce rotating fields of quantum fluctuation dipoles that allow rotating bodies to push against the vacuum. Likewise, all bodies and particles push against quantum fluctuations.

The mechanical force is apparent on much larger scales then simple tops. The linear mechanical force between bodies of matter causes bodies of matter to accelerate away from each other. The rotation of the sun causes elliptical orbits of planets to precess. Mechanical Lorentz forces on stars in rotating galaxies cause those stars to form into spirals. Even inside stars and planets, rotation causes counter-rotation that leads to dynamo effects. The mechanical field force explains most phenomena popularly attributed to the curved space model.

The curved space model runs into serious difficulties when you consider more than two objects in very different orbits. For example, the Earth has a nearly circular orbit, but a comet passing near the Earth is following a large elliptical path that is nearly perpendicular to the Earth's orbit. How can space be simultaneously curved in two perpendicular directions? It makes no sense. If one considers millions of photons crossing a single point in space from different directions, it makes even less sense.

It makes no sense to adopt a curved space model when a geometrically flat space model works. Given space has no physical dimensions of its own it can only be thought of as flat. Why do physicists pretend non-physical space has physical dimensions? Probably they do it because science fiction is more fun and interesting to think about than real science. Any true scientist would insist on flat space geometry over curved space, and find suitable flat space theories. Oh, and by the way, the uniformity of the CMB is evidence that the aether rest frame is geometrically flat.

Curved space is definitely one of the greatest lies in physics. It was a huge unnecessary step backward in physicists' understanding of the universe.

[63] P. Davies, The Goldilocks Enigma, First Mariner Books, p. 43, 2006

Lie #44: General Relativity

The 44th of the greatest lies in physics is general relativity. After the last few lies, this lie was obviously coming. As with so many other great lies of physics, it is largely due to the denial of zero-point energy. For if zero-point energy is considered, general relativity fails.

Physicists also failed to recognize earlier that even the simplest mechanical force interactions such as inertia and the gravity opposing force of a spinning top, require that there be a fundamental mechanical force that includes fields of quantum fluctuations. Here is a brief list of some of the most significant failures of the general relativity model.

   A. It is incompatible with the known existence of aether, as the aether rest frame has uniform physical and time dimensions as derived from quantum fluctuations and is geometrically flat.
   B. It does not explain how information about the mass and other forms of energy are transmitted instantaneously throughout the universe.
   C. It does not explain how bodies are accelerated.
   D. There is far too little matter in the universe for the general relativity model to be valid, the so-called missing matter or dark matter problem.
   E. It does not explain physical space curvature.
   F. It does not explain how space is curved differently at the same point and time for bodies or photons, moving at different velocities and/or directions.
   G. If zero-point energy were included, without changing coefficients of the general relativity model, the universe would collapse to the smallest possible volume.
   H. If zero-point energy were included in the general relativity model, and the coefficients adjusted, while assuming constant zero-point energy, the gravitational effects would be negligible since non-zero-point energy is an insignificant percentage of the total.

I. The total energy of space is constant, after recognizing that mass and other forms of energy are equal to the zero-point energy they exclude, so there is no variation in total energy density and no basis for gravity being due to changes in local energy density.
J. When zero-point energy is factored in, there is not enough mass-energy to curve space even if space were physical and could be curved.
K. A fundamental mechanical force more easily explains most so-called proofs of general relativity, and even more related phenomena not explained by general relativity.
L. A flat space model is required to describe the dimensions of the aether rest frame.
M. The uniformity of the CMB is scientific evidence that the aether rest frame is geometrically flat.
N. It is impossible to combine the dark force, high-energy gravitational force, and the mechanical force into a single curved space model.
O. High-energy gravity is best accounted for by the Fatio-Casimir force, which must exist anyway.
P. It is impossible that both the Fatio-Casimir force and general relativity to exist.

There is additionally the problem that stable matter locally increases the van der Waals torque which increases the permittivity and permeability while decreasing the speed of light. The increase in permittivity and permeability near the Sun for example, changes the dielectric constant and speed of light causing light to deflect. The deflection of light and related phenomena are variable speed of light problems rather than curvature of space problems. This leads to a form of variable speed of light relativity that replaces those parts of general relativity.

General relativity truly is one of the greatest lies in physics.

Lie #45: Gravitational Time Dilation

The 45th of the greatest lies in physics is gravitational time dilation. Here is a concept that physicists' screw up on multiple levels due to numerous misconceptions and lies. What we do know is that clocks run slower near bodies of matter, specifically the Earth, as this has been experimentally proven.

The largest body of proof is from clock rates of clocks in orbit, particularly those used in the global positioning system. GPS clocks run faster due to their position relative to the Earth, and somewhat slower due to their velocity, with the net difference being that they run faster than Earth based clocks while still on Earth's surface.

So some of you are probably thinking, how is this a lie if it has been proven? The lie comes about because mainstream physicists have not actually identified the physical mechanism responsible for clock slowing. The idea that clock slowing is related to gravity and some sort of dilation of space are merely bad guesses which taken together form a lie. Space by itself does not have physical clocks, it contains physical clocks due to the frequencies of the quantum fluctuations of the quantum field.

As mentioned in Lie #16, time dilation of space is a lie with respect to special relativity theory. **Time dilation of space is also a lie with respect to the general relativity model or any other gravitational model.** In addition to the arguments in Lie #16, it is also impossible to have space-time dilation due to two opposite but approximately equal forces.

The scientific evidence does show that clocks slow near bodies of matter, so we must forget about ideas of time dilation due to space, and focus of mechanisms that can cause clock slowing. To understand how clocks slow, we need to understand how matter affects the universal clock rate, and more basically, what is the universal clock rate, and how does it come about.

In order to actually solve the puzzle of how clock rates change we need to know what physical thing regulates the universal clock rate and what slows that rate near bodies of matter. First, the standard clocks are the quantum fluctuations in the aether rest frame. The universal clock rate, which is the fastest possible clock rate—not in a Casimir cavity[64]—is achieved when a clock is at rest with respect to the aether rest frame, assuming there is no matter in the vicinity. The clock rate must be regulated by the same mechanism that regulates the quantum fluctuations.

From relativity theory, as discussed in Lie #16, clock rates slow when clocks move at a velocity relative to the universal rest frame. This tells us that the clock rate change must be due to interactions with quantum fluctuations. Or to turn it around, quantum fluctuations regulate clock rates. Here we have to note that a quantum fluctuation rotates with a maximum wavelength (distance) and frequency (time) related to the speed of light, $c = \lambda \nu$. The speed of light also slows in moving rest frames and near bodies of matter to the same degree as the slowing clock rate.

The rotation of quantum fluctuations slows due to an increase in the van der Waals torque of the quantum field induced by the surrounding field of quantum fluctuations. The van der Waals torque is greater when measured in a reference frame moving with respect to the rest frame. The torque is also greater in the vicinity of a body of matter. When the van der Waals torque increases, clocks slow.

Gravitational time dilation is a lie. Clock slowing is not directly due to gravity and space-time does not get dilated. Van der Waals torque changes due to the local distribution of matter which changes the local clock rates.

[64] Inside a Casimir cavity some vacuum fluctuation wavelengths are excluded. The exclusion of wavelengths leads to a reduction in the van der Waals torque of the vacuum. It was proposed by Klaus Scharnhorst that the speed of light is faster in a Casimir cavity.[65,66] If so, it is also true that clock rates are faster in a Casimir cavity. The change in speed of light and clock rate are proportional.

[65] G. Barton, K. Scharnhorst (1993). "QED between parallel mirrors: light signals faster than c, or amplified by the vacuum". Journal of Physics A. 26 (8): 2037.

[66] K. Scharnhorst, "The velocities of light in modified QED vacua". Annalen der Physik. 7 (7–8): 700–709, 1998. arXiv:hep-th/9810221.

Lie #46: Gravitational Time Dilation is a Proof of General Relativity

The 46th of the greatest lies in physics is gravitational time dilation is a proof of general relativity. Of course, if general relativity is a lie, this statement must be false. This claim is, however, repeated over and over by mainstream supporters of the general relativity model, so it deserves a place in the top 100 of physics follies.

Physicists never once tell us how physical clock rates change due to the general relativity model. At best we are given a vague argument that clocks slow when they are accelerated, without stating how acceleration slows the clock works. Or worse, they say that space has its own clock that is somehow slowed due to gravity, the mythical time dilation of space. They never say how non-physical space has physical clocks. The only important piece of information we got out of this is that mechanical and gravitational forces change clock rates the same way.

The statement that gravitational time dilation is a proof of general relativity is a lie. Based on lie #45 it is easy to see that clock rates have nothing to do with gravity, at least not directly, and there is no dilation going on. Clock rates are related to the frequencies of quantum fluctuations under local conditions, with regard to the proximity of matter and velocity relative to the aether rest frame.

Lie #47: The Equivalence Principle Proves General Relativity

The 47th of the greatest lies in physics is the equivalence principle proves general relativity. On the day of this writing the equivalence principle page of that bastion of obsolete physics, Wikipedia, began:

> In the physics of general relativity, the equivalence principle is any of several related concepts dealing with the equivalence of gravitational and inertial mass,"[...]"Einstein suggested that it should be elevated to the status of a general principle, which he called the "principle of equivalence," when constructing his theory of relativity:...

Here is a case where Einstein was correct. The equivalence principle is a general principle. The equivalence principle is true even though general relativity is a lie.

In the guise of the equivalence principle, inertial mass is thought of as a pseudo-force rather than a real force, and this pseudo-force is said to be equivalent to a real force, gravity. The obvious mistake here is that they are both real forces. The inertial force is a real mechanical force, not a pseudo-force. The equivalence principle can be simplified by just saying a force is a force.

The equivalence principle does not actually provide a mechanism for gravitational interactions between bodies of matter or photons, so it does not actually solve the related physics problems. This is another rule that ignores the underlying physical mechanism, the actual physics. It does, however, give us a better way to look at certain problems if we flip it around.

Instead of saying that inertia is equivalent to gravity, we should be saying that gravity is equivalent to inertia. The gravitational force is equivalent to the combined electro-mechanical forces that aether exerts on a body. Or, even more generally, gravity is not even a force, as it is the electromagnetic and mechanical forces that are fundamental. And because the mechanical forces apply to electromagnetic forces on electrically charged bodies the generalized electromagnetic force unifies electromagnetic, mechanical, and gravitational forces. The change in the permittivity, permeability, and speed of light in the proximity of matter is also an electromagnetic phenomenon.

The statement that the equivalence principle proves relativity is a lie. Einstein used his interpretation of the equivalence principle to support his theory of general relativity, while the equivalence principle is actually a separate higher-level principle. A force is a force.

Lie #48: Dark Matter

> *The fact that the dynamics of galaxies and clusters of galaxies do not match standard gravitational theory with the observed or luminous matter can in principle be explained in 2 different ways: we are using the wrong theory of gravity, or we are not seeing all the matter. The latter is the more common view. However, there is no clear consensus about what the dark matter may be.*[67]
>
> Paul S. Wesson

The 48th of the greatest lies in physics is dark matter. Dark matter is thought to be matter in the universe other than normal matter made of protons, neutrons, and electrons. The reason physicists believe in dark matter is that 70% to 90% of matter needed for the popular gravitational models to be correct is missing. This shortage of matter was discovered in spiral galaxies and galactic clusters, where the velocities of stars and galaxies respectively were too great for gravity alone to keep them together given the amount of mass present. Hence this is called the missing mass, missing dark matter, or just the dark matter problem, and it is often cited as one of the top ten unsolved problems in physics.

Physicists still want to believe that the gravitational models are correct, even though they appear to be 90% wrong. It is sort of like getting 10% of the questions right on an exam and receiving a 100% grade. This view is baffling to anyone with any common sense. It is one thing to hold on to a theory that needs incremental improvements, and another thing to support one that is almost entirely wrong.

The correct conclusion all along should have been that physicists are using the wrong theory of gravity, or that there are additional, presently unaccounted for forces at work. This should be even clearer now that we know that the popular gravitational theories are actually trying to model a combination of two or three forces. It is also clear due to the universe's black hole problem where the entire visible universe would form a black hole if the gravitational constant is constant, that G is not a constant on the scale of the visible universe.

The problem with understanding the formation of spiral galaxies is readily fixed once we factor in the mechanical force. The tangential velocity of stars in a rotating galaxy leads to a mechanical Lorentz force pushing fast moving stars toward the center of the galaxy. Without this Lorentz force the outermost stars in a spiral galaxy would fly off into intergalactic space. This force also solves the galactic cluster velocity problem.

In basic electromagnetic forces it is known that two parallel wires carrying currents in the same direction are attracted toward each other. The same is true for two stars on a parallel path. That is how stars form bands in rotating galaxies. Without the mechanical force, galactic spirals would not exist. There was never a dark matter problem; it has always been a missing force problem.

The idea of dark matter is a lie. It was the popular gravitational models that were entirely wrong.

[67] P. Wesson, "Fundamental Unsolved problems in physics and astrophysics," prepared for California Institute for Physics and Astrophysics, http://www.calphysics.org/problems.pdf

Lie #49: Gravitons

The 49th of the greatest lies in physics is gravitons. As mentioned previously gravitons are hypothetical elementary particles that are said to be the gauge bosons for gravity. In other words, it is a smart particle tasked with transferring information about gravity from particle to particle, and body to body. As discussed in Lie #18, all gauge boson models are a lie due to numerous inconsistencies with physical reality, and gravitons are no different.

Gravitons are a way to shape Newtonian gravitational theory into something similar to electromagnetic theory. In the electromagnetic case, photons are said to be the gauge boson or force carrier (Lie #7). In order to make gravitational theory more like electromagnetic theory, physicists invented the graviton hypothesis. Note that Einstein's general relativity model does not rely on a graviton, since space curves due to some secret instantaneous handshake with all the matter and energy in the universe, rather than having bodies move in response to a force.

Even though gravitons never made it past the hypothetical stage, and have not been detected directly or indirectly, they are still included in the standard model tables of elementary particles. Gravitons should never have made it that far.

In true quantum gravity, as described by Fatio and Casimir, quantum fluctuations are responsible for the quantum pressure that causes gravity and gives bodies a push to accelerate them. Fatio-Casimir quantum gravity is entirely consistent with, and part of electro-mechanical force theory. No gravitons are required.

Gravitons are a lie. Gravitons do not exist, and never should have been made part of any theory.

Lie #50: Gravity is Due to all Forms of Energy

The 50th of the greatest lies in physics is gravity is due to all forms of energy. As part of his theory of general relativity, Einstein assumed that all forms of energy gravitate, or cause gravity. This bad assumption is partly responsible for the slow acceptance of modern aether theory. As discussed before, if zero-point energy gravitates, under the general relativity model, the universe would be compressed to a black hole.

Einstein made this assumption since Poincaré, Pietro Olinto, and later Einstein himself showed that matter and energy are equivalent by the formula $E = mc^2$, or as Poincaré published it $m = E/c^2$.[68,69,70] That said, others had considered the relationship prior to Poincaré. Newton's theory of gravity states that gravity is proportional to mass. So the obvious extension of Poincaré's finding was that since energy is equivalent to mass, energy must lead to gravity too.

This led to the question; does zero-point energy gravitate or does it not? If it does gravitate, as required by the general relativity model, then it invalidates the model. If it does not gravitate, it invalidates Einstein's assumption. And, if it gravitates in a way that is inconsistent with general relativity, it invalidates the model. No matter how you slice it, zero-point energy is a big problem for the general relativity model. So Einstein and his followers became big aether deniers, if they were not already.

Based on the Fatio-Casimir effect, zero-point energy does gravitate. It gravitates through van der Waals forces pressing on and pushing bodies. The reason this does not have disastrous consequences is because the pressure is almost perfectly equal from all directions. That, and particles do not collapse under this pressure.

It is only small pressure differentials that lead to motion resulting in the force we call gravity. Gravity is not really a separate force at all, but just another outcome due to van der Waals forces produced by quantum fluctuations, a part of electromagnetic and mechanical force theory.

The assumption that gravity is due to all forms of energy is a lie. At least it is a lie in the context of the general relativity model. The question of what causes gravity is addressed in the next lie.

[68] J.H. Poincaré, Arch. neerland. sci., 2, 5,232, 1900, J.H. Poincaré, In Boscha, 1900:252.
[69] O. De Pretto, "Ipotesi dell'etere nella vita dell'universo (Hypothesis of Aether in the Life of the Universe)." "Reale Istituto Veneto di Scienze, Lettere ed Arti," (The Royal Veneto Institute of Science, Letters and Arts) LXIII (II): 439–500. (accepted November 23, 1903 and printed February 27, 1904).
[70] A. Einstein, "Zur Elektrodynamik bewegter Korper." Annalen der Physik 17:891, 1905.

Lie #51: Gravity is Due to Mass

The 51st of the greatest lies in physics is gravity is due to mass. It may be shocking that anyone would call into question the theory accepted since Newton's day that mass causes gravity, but the existence of the Fatio-Casimir effect reopens the question. If we think about it, there has never been a successful answer to the question of exactly how mass causes gravity. So, perhaps it has never been due to mass at all.

The mass-gravity relationship appears to work very well, but we often forget that masses of bodies are derived in a somewhat circular fashion. We determine the mass of Jupiter, for example, by measuring its orbit and calculating its mass based on Newton's equation and then we say it has that mass. We do not know the mass of Jupiter by any independent means as there is no scale or balance out in space we can use. Does Saturn truly have half the mass density of Jupiter, or is that an artifact due to an additional force missing from gravitational theory. Presently we trust that Newton's formula is correct and move along.

We never think that since Newton and his followers do not know how mass causes gravity, perhaps the masses of Jupiter or Saturn are somewhat different, and the actual cause of gravity may be something different. That said, the Newtonian model works so well, at least within the solar system, mass must be a good approximation for whatever really causes gravity.

We can think about plausible physical explanations for gravity by considering the Fatio-Casimir effect more carefully. Fatio recognized that when a body is pushed, that force must not be perfectly transmitted to the other side of the body or the shadow effect would not exist. That means quantum field energy pushes on particles, but the particles in turn do not push equally on the quantum field on the opposite side. This is an inelastic response as some of the energy is somehow getting lost.

It is this loss of energy that leads to the shadowing effect, which leads to gravity.

As a brief aside, Fatio also recognized that bodies must be mostly empty space or his proposed force would be way too strong. He was perhaps the first to recognize that bodies of matter are mostly empty space.

If we think about this from the perspective of the particle, the van der Waals forces of the quantum fluctuations are able to push on the particle but the particle is unable, or less able, to push back. It is this difference that causes gravity at the particle level. This also tells us a little something about particle structure, but that is a very deep rabbit hole we will not go down in this book.

How then does this structural differential relate to mass? The answer comes down to size. Smaller quantum fluctuations are more energetic, and when van der Waals forces are pushing on particles, that push comes from quantum fluctuations that are approximately the same size as the particles they are pushing on.

The greatest gravitational force is then on the smallest particles, the protons and neutrons. The push on the larger electrons is much less great, so electrons have less effect in a body made of roughly the same number of protons and electrons.

This relationship is based on Dirac's idea of mass-energy, as confirmed by my calculations; that a particle's mass is related to its diameter and the quantum fluctuation wavelengths that the particle displaces. Mass is effectively another measurement of size when it comes to the permanently stable particles.

What about other forms of mass or energy? It all comes down to if and how they impede the transmission of van der Waals forces. If a quantum fluctuation is rotating as part of a moving body's inertial field, and thus is effectively inertial or relativistic mass, then it is less likely to respond to other dipoles in typical van der Waals interactions.

In the end, all forms of non-zero-point energy and mass-energy are due to zero-point dipoles being involved in whatever process produces or stores that energy. Once the dipoles have another function, their ability to transmit van der Waals forces is diminished.

Saying that gravity is due to mass is a lie. Gravity is due to particles' inelastic response to van der Waals forces between quantum fluctuations, which leads to the Fatio-Casimir effect. The strength of those forces is related to particle size, which is then tied to mass. Mass and gravity are entirely electromagnetic. Other quantum fluctuations tied to a process that produces or stores energy in the form of a field of quantum fluctuations, also inhibit the transmission of van der Waals forces, leading to a gravitational effect.

All that said, mass will likely remain as the simplest way to approximate gravitational forces for engineering purposes.

Lie #52: The Pauli Exclusion Principle

The 52nd of the greatest lies in physics is the Pauli exclusion principle. The Pauli exclusion principle is the principle in quantum mechanics that two identical fermions—½ integer spin particles such as electrons—cannot occupy the same quantum state simultaneously. Note that this principle does not apply to bosons such as photons. The lie in this case is not due to the underlying concept of two particles not occupying the same state, but that it fails to approach physics from the perspective of explaining physical phenomena.

The trouble began with spin, as spin is a non-physical property, and it is said to be an intrinsic property of particles. There has never been a standard physical model for spin. On the other hand, when physicists agree on a particle structural model, spin will likely end up being spin of some sort since rotation is required to produce a magnetic field. The exclusion principle and spin were the start of physicists adopting an arbitrary rule-based approach to particle properties, rather than basing properties on physical phenomena. In other words, they are not performing real physics.

The two electrons orbiting a helium nucleus, for example, are said to have two different spin states, so they are in different quantum states. It is known that these states do lead to different transition energies, so there is something physical going on. Then this rule-based approach to understanding this difference in energy states is expanded into a broader spin theory.

Note that the first three quantum numbers in atomic theory are the principal quantum number, azimuthal or angular momentum quantum number, and the magnetic quantum number. Each is based on a physical property. The principle quantum number, from the Bohr model perspective for an electron, is the shell being occupied, essentially related to the distance from the nucleus. This distance is said to be quantized, as only certain distances or energy states can be occupied. The particle's angular momentum and magnetic field are also based on those real physical properties. Electric charge is the other quantum number that is generally treated as an intrinsic property, but probably is a physical property, more on that in the next chapter.

The spin quantum property is the first failure of the Pauli exclusion principle, but the principle has failed many more times since. Each time this principle fails, a new quantum property is invented to correct the failure, and each of these new make-believe quantum properties has no physical basis. These additional quantum numbers are now more commonly called flavors.

Other flavors include weak isospin, lepton number, baryon number, hypercharge, strangeness, topness, and bottomness. Then these and other fictional intrinsic quantum properties have been joined together to form the basis of a numerological particle theory—quantum chromodynamics—also known as quark theory. Unfortunately, there is no physical explanation for any of these properties. Particle physicists simply forget that they are supposed to be producing physical explanations. They forget they are physicists.

The Pauli exclusion principle is one of physics' rule-based principles that has no basis in physical reality, and does not truly deserve to be called a principle. Particles do not 'know' the rule. This principle is an example of the dangers in coming up with arbitrary rules and calling them principles, instead of coming up with physical solutions to the underlying problems.

The Pauli exclusion principle is a lie. It is better stated as a test to tell us when to look for physical differences between otherwise identical looking particles or resonances.

Lie #53 Charge is Intrinsic

The 53rd of the greatest lies in physics is charge is intrinsic. Charge is one of the most important particle properties. The best-understood and most fundamental force, electromagnetism, is due to electric charge. A static charge causes dipoles in space to be polarized and form an electric field. Moving charges cause dipoles in space to rotate, forming a magnetic field. Dipoles only exist because there is charge, and just as importantly two types of charge, which we define as positive and negative.

As important as charge is to our understanding of the universe, standard model physicists have no idea what charge is. They have no idea how to physically explain what charge is. What is charge? They do not know. How do particles have charge? They do not know. How are there positive and negative charges? They do not know that either.

Because physicists are ignorant of charge, they punt, and treat charge as an intrinsic particle property. It just is. Even Dirac, when he formulated the equation that bears his name, treated charge as an intrinsic property, while in a complete equation of an electron both charge and mass should be described more fundamentally.

Like with mass (Lie #19), treating electric charge as an intrinsic property is an abdication of responsibility. Physicists are tasked with finding a deeper understanding of physical properties through physical understanding and mathematics. Charge is the basis for electromagnetic theory, so if we do not understand charge, we do not truly understand electromagnetic theory.

Fortunately, while there are 122+ masses that need to be explained, there are only two charges, positive and negative, with the same magnitude, assuming we disregard the fractional charges in quark theory. We should think that two charges would be far simpler to deal with.

One aspect that makes the charge puzzle challenging, is that the two stable particles, the electron and proton, have equal but opposite charge, and yet they are physically different sizes and different masses. And then, their antimatter opposites also have the same magnitudes of charge but opposite. Charge is either independent of size, or it is a solution tied to a specific structural problem, such that only these two sizes are stable, somehow leading to identical magnitudes of charge.

To avoid these problems, and the task of providing any physical explanations at all, physicists have simply decided that all particles are point particles and those point particles have a point charge. **Point charge is another lie within the lie of intrinsic charge.**

The problem with the concept of a point charge is infinity. We know that a particle's charge has a fixed value. If the particle is a point, in other words infinitely small, then the charge density is infinite, assuming charge is a distributed property. As with mass, infinite charge density is not something we can model. It cannot be physically real.

We can turn that around. If we do have a physical model for charge composed of smaller units of charge, and we take that model to an infinitely small size, we end up with one of two conclusions. The charge will be zero, or it will be infinitely large due to a divide by zero error.

There have been attempts to physically model the charge of particles with physical dimensions. The most common approach is to have charge evenly distributed over a surface or volume in an infinitely smooth manner, or with numerous points of smaller units of charge. This is the charge is equal to the sum of a bunch of smaller charges approach. These approaches also do not address the question of what charge is or how charge can be physically explained. They basically say that charge is intrinsic but at a smaller scale.

In order to be good physicists and find a physical explanation for charge we have to get past the 'charge is charge' concepts, and look instead for something that is not charge, but rather a physical phenomenon that causes attraction and repulsion mimicking charge-like interactions. Charge as a form of positive and negative charge-energy is one possible approach. For now, however, there is no commonly accepted explanation for how charge works.

In-between publication of the first and second edition of this book I realized that electrons and protons have the same magnitude of charge because the magnitude is due to the polarizability of the quantum field.[71] The electron and proton act like simple polarizers, polarizing quantum dipoles made of simple polarizers. Each simple polarizer induces the same amount of polarization in the quantum field and that is where the charge magnitude comes from. That is how different particles come to have the same magnitude of charge.

To say that charge is intrinsic is a lie. The reinterpretation of Gauss' Law explains the charge magnitude and tells us that the magnitude of electric charge is not intrinsic. It reduces the charge question to how does a particle become a polarizer that can polarize or be polarized? As good physicists we must still endeavor to explain polarization and not treat it as an intrinsic property.

At their most fundamental level electrons and protons are electrical polarizers that are also matter or antimatter. Those two properties appear to be the two most elementary particle properties. It may be that we can never explain them but we must try.

[71] R. Fleming, "Electron Properties Explained as Quantum Field Effects." GSJournal.net, 18 Sept. 2018.

Lie #54 Point Particles

The 54th of the greatest lies in physics is point particles. The point particle problem comes about due to what can only be described as intellectual laziness. Instead of trying to tackle a physical description of particles, physicists instead assume that particles exist at a point and that all their properties are intrinsic.

Intrinsic point particle properties problems were addressed in Lies #19, #20, and #53 regarding mass and charge. The basic problem with point mass or point charge is infinity. A particle's mass and charge are fixed at well-known values. If the particle were a point, in other words infinitely small, then the mass or charge density would be infinitely large, assuming that mass and charge are properties that can be distributed. Infinite mass or charge density cannot be physically real. Distributed smaller units of intrinsic mass or charge cannot be physically real either.

Right off the bat, mass is a deal breaker for the point particle model. Dirac hypothesized that the mass of the electron and positron is due to the energy needed to maintain a hole in the Dirac Sea, the aether. This has been shown to be correct, not only for the electrons, but for protons and neutrons based on their experimentally verifiable physical dimensions. The masses of resonances are somewhat different, but we will see they are also ultimately due to the exclusion of zero-point energy, but in the form of relativistic mass.

If we imagine a particle much smaller than its known radius, it would exclude smaller more energetic quantum fluctuations, and thus its mass would be much larger. If a particle were the size of the Plank length it would effectively be a small black hole with a huge mass. Once you get to a true point mass, it must either be infinitely large or zero, and neither of those answers are correct.

Electric charge has the same problem as pointed out in the last chapter. A point charge will be zero, or it will be infinitely large due to a divide by zero error. Particles with distributed charge cannot be point particles, and distributed charge cannot explain what charge is. Charge must be modeled a different way. The charge magnitude is ultimately due to the amount of polarization of the quantum field due to a single polarizing particle. But that still leaves the problems of physical structure and how a polarizer polarizes.

Matter and antimatter represent a problem much like electric charge, as the Dirac equation shows that they are positive and negative 'energy' solutions. In a point particle the 'energy' that determines the matter or antimatter character of the particle must be zero or infinite. Like electric charge, matter and antimatter must be modeled a different way. The amount of matter and antimatter can be determined the same way as the amount of electric charge, as an effect due to the polarization of the quantum field due to matter polarization. The question of how a particle polarizes with respect to matter and antimatter is still open.

Particles also have angular momentum, a magnetic field, and spin. All three of those properties require that particles have physical dimensions in order for us to successfully model them. A point does not have angular momentum. A point cannot produce a magnetic field. Rotation is needed to produce a magnetic field and that rotation is related to spin. Real physical spin requires that particles have physical dimensions.

The thing is that most physicists understand these problems, and yet they keep repeating the point particle lie. It truly seems to be due to intellectual laziness along with the inability to admit that they simply do not know the structure of particles. Physicists apparently would rather tell a lie than say, "we do not know," or to acknowledge new theories.

Lie #55: Small Electrons

The 55th of the greatest lies in physics is small electrons. Small refers to theories that place the size of an electron variously from a point particle, the size of the Planck wavelength, 10^{-22} meters, 10^{-18} meters, 10^{-15} meters or any other physical diameter smaller than 10^{-12} meters. The radius of the electron is unknown because the structure of the electron is unknown to standard model theorists; although some physicists may go as far as to say it does not have structure. In general, the physics mainstream thinks that an electron is smaller than 10^{-12} meters, which is the size equating to the Compton wavelength.

Much of the confusion comes about because of suppositions about electron structure. When attempting to model electrons, physicists often assume that the electron mass is due to a collection of smaller masses, and a unit charge is due to a collection of smaller charges. And then the magnetic moment is assumed to be due to the motion of the smaller charges. Spin and other properties are usually left aside while physicists focus on these basic properties. Distributed masses and charges are then thought to be organized in some way. A multitude of possible solid geometric objects, surfaces, or lines have been tried, starting with the most obvious spherical or spherical shell models.

One of the most basic arguments against a large electron is that if an electron is made of distributed charges, and the magnetic moment is due to their motion, then the outer surface of the electron would need to exceed the speed of light. This model is unworkable if we assume the speed of light truly limits the rotational velocity as it must. As is often the case, physicists overlooked the obvious alternative answer. Electron charge magnitude is due to the polarized quantum fluctuations.

If instead of using electron mass to calculate the magnetic moment we use the electron's Compton wavelength, we find that the magnetic moment is due to a quantum fluctuation structure with the diameter of the Compton wavelength, the electron's charge, and rotating at the speed of light. This is the nature of the electron's external quantum field structure.[71]

Similarly, electron mass is not due to the sum of smaller masses. As Dirac predicted the mass of the electron and positron are due to the energy required for them to exist in the aether. Those particles displace a certain amount of zero-point energy. For Dirac's hypothesis to be correct, the diameter of the electron must be the electron's Compton wavelength. Based on Dirac's model of electron mass, smaller electrons would be far too massive as discussed in Lie #19.

Unfortunately, physicists have not been performing photon or electron scattering experiments to officially determine the size of a Compton sized electron, so a precise value is unknown. The above experiments are not performed because physicists decided that high-energy scattering experiments would be more precise. But instead of being Compton sized, the high-energy scattering experiments result in much smaller measurements.

Physicists chose not to reconcile these two vastly different measurements, but on a strictly ad hoc basis decided that the high-energy scattering experiments were correct. They failed to recognize that there are two interpretations to the high-energy scattering experiments; one is that the electrons are small, and the other is that the electrons are transparent to high-energy particles. Physicists disregarded the second, more obvious answer, which allows for a viable explanation for mass.

Another part of the problem is that if an electron is large, on the order of the Compton wavelength, and transparent to high-energy particles, that implies that it has structure. And, if an electron has structure then perhaps it is not an elementary particle, since whatever it is made of must be elementary. Never mind that whatever it is made of may not exist except while in the structure of an electron or other particle. The same argument is made about protons, that they must not be elementary since they have detectable structure.

The quantum fluctuations that inhabit and electron's interior and outer shell, definitely have a separate existence. Electrons appear to have a small, perhaps even point-like electrical and matter-antimatter polarizer surrounded by quantum fluctuations. These stable bare electrons cannot exist in free space without their external quantum field structure. It is likely however that, because they are so short-lived, quantum electron-positron pairs should be thought of as bare electron-positron pairs without the external structure.

The small electron theories are lies. The path forward is to learn more about the physical structure of Compton sized electrons through scattering experiments, since electron mass and magnetic moment are physical properties due to that physical dimension.

Lie #56: There is No Repulsive Force Between Electrons and Protons

The 56th of the greatest lies in physics is there is no repulsive force between electrons and protons. Within the standard model of physics, the only force between electrons and protons is coulomb attraction. This is the basic electrical attraction due to the proton having a positive charge and the electron having a negative charge.

There is no repulsive force between electrons and protons in the standard model. But, if there truly were no repulsive force between electrons and protons, electrons would fall into protons due to Coulomb attraction and form neutrons. There would be no hydrogen in the universe.

Experimentally we know that when electrons get close to protons there is a repulsive force that has been measured with energy of approximately 780 keV (kilo-electron-volts). For comparison, the mass-energy of an electron is 511 keV, so this potential barrier represents a tremendous amount of energy at the particle scale. This is not something negligible or something to be readily dismissed, and yet standard model physicists ignore it when considering force theories.

This potential barrier energy prevents electrons from simply falling into protons and yet physicists make no attempt to determine the nature of this force. Instead they use a wave equation to describe the energy states of an atom and neglect that there is a force that can balance their equations.

So now we must amend our list of the fundamental forces. They are the:

1. mechanical force,
2. electromagnetic force,

3. weak force,
4. strong force,
5. dark force,
6. high-energy gravity force,
7. repulsive force between electrons and protons.

The list of fundamental forces is starting to get too complicated, so there must be an easy way to simplify it. The repulsive force between electrons and protons is a repulsive force between two particles of matter, so that implies a force that causes matter to be repelled from matter. The dark force is also a force that causes matter to be repelled by matter. Additionally, the mechanical force also causes matter to be repelled by matter.

Conversely matter would be attracted to antimatter at very close distances. It would be interesting to shoot positrons at protons and see if this is a measurable effect. However, the Coulomb repulsion between those two positively charged particles dominates the interaction at normal distances.

The lie that there is no repulsive force between electrons and protons is truly one of the greatest lies of physics. It is a lie that has severely limited physics development over the past 90 years as physicists continually failed to recognize that a force was missing from their theories.

Lie #57: The Schrödinger Equation Describes Electron Motion

> *A consequence of the zero-point energy was that an electron in a stationary atomic state does not spiral into the nucleus if it absorbs zero-point energy at a rate which balances the rate at which it radiates energy. Only if the latter exceeds the former does it radiate. What zero-point radiation really does is to permit the concept of ground state, from which further decays do not occur.*[72]
>
> Peter Finley Browne, 1995

The 57th of the greatest lies in physics is the Schrödinger equation describes electron motion. The Schrödinger equation is a wave equation developed by Erwin Schrödinger in order to explain the energy states of electrons around the nucleus of a hydrogen atom.[73] The development was aimed at fixing difficulties with the Bohr model. It has since been extended to other systems and is widely considered one of the most important developments in quantum mechanics.

The Schrödinger equation gives a more fundamental explanation for the quantization of electron states, commonly called orbits, although they should not be thought of as orbits in the classical sense, but rather energy states. It was also thought that the wave equation solved the problem of how an electron does not radiate photons as it changes directions in its orbit, as electrons are thought to do in classical theory.

While the Schrödinger equation is no doubt highly successful, it fails at providing answers to certain basic questions about electron motion.

    A. It does not explain how an electron moving directly toward a proton, accelerated by the electric force, does not fall into the proton.

B. It does not explain how an electron moving directly toward a proton would change direction and start to 'orbit' the proton in a manner consistent with the wave equation.
C. It does not incorporate the repulsive force between the electron and proton.
D. It does not actually explain how the electron does not radiate photons as it moves, in terms of providing a physical mechanism.
E. Electrons are known to occupy a quantum state that looks like a cloud, while jumping from place to place within the cloud. The equation does not describe this jumping motion.
F. Each electron cloud includes a range of distances from the nucleus, so in a classical sense, each jump includes a small change in energy while not radiating. This is not explained.
G. Electrons jump from one orbital state to another without accelerating, decelerating, or occupying the intervening space, in violation of classical principles of physics. This is not explained.
H. These orbital jumps are accompanied by the emission or absorption of a photon, depending on whether the energy state is decreasing or increasing. These photon interactions are not explained by the wave equation. Note that this will be examined in the next chapter.
I. It does not explain, in a physical sense, how electrons do not occupy identical quantum states. The Pauli exclusion principle is after all, a rule-based approach, which does not include a physical mechanism. As such, it is incomplete.

The first three A, B and C, tell us that there is a need for a force balance equation that includes the ~780 keV repulsive potential between the proton and electron. If this force could be incorporated into a wave equation, that would be even better. D through H tell us that there is some physical mechanism missing from our quantum models, and I points to a different physical mechanism or force missing from our models.

The questions in D to H are fortunately easy to explain once we recognize that the space inside an atom is still filled with quantum electron-positron pairs, and quantum positrons can and do annihilate with free electrons. So when a quantum positron gets too close to a free electron, they annihilate each other. That annihilation leaves the quantum electron free to inherit the quantum state of the free electron that was destroyed. This mechanism is sort of like Hawking radiation but instead of a virtual particle being captured by a black hole, it is captured by annihilation. This is illustrated in figure 57-1.[74]

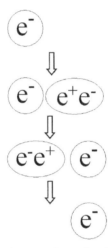

Figure 57-1 A basic quantum fluctuation mediated annihilation-production event wherein a free electron is annihilated and the quantum electron takes its place.

This type of annihilation-production mechanism is simple and elegant, and importantly there is nothing prohibiting it from happening. We can confidently state that such interactions must occur.

These interactions give us a clear physical model for how electrons jump within an electron cloud. The minor energy variations within a single quantum state are then accounted for by small amounts of energy coming from or being returned to the zero-point field. In this respect an electron can be thought of as being in balance with the zero-point fields as stated by Browne in the quote above, which was an idea with origins in the work of Planck and Nernst.[72] That takes care of D through F.

As for orbital transitions, we can see how annihilation-production interactions work there, as illustrated in Figure 57-2.[74] But in this case, there is a change in the energy state of the electron, so a photon must be involved in the interaction. If the electron jumps to a lower-energy, lower orbital state, a photon is produced emitting that amount of energy. In order for an electron to jump to a higher-energy state it must absorb a photon. That takes care of G and part of H.

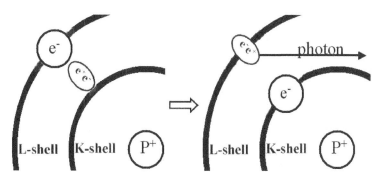

Figure 57-2 A Bohr style hydrogen model showing an electron-positron quantum fluctuation mediated electron orbital transition with the emission of a photon.

That is only a small sample of the types of interactions that are mediated through annihilation-production interactions with quantum fluctuation dipoles. All quantum jump type motion on the quantum scale ultimately happens this way. Every particle in our bodies is continuously being annihilated and reproduced as we move through space. For more in-depth discussion, please see my paper "Quantum jumps as vacuum fluctuation particle pair interactions analogous to Hawking radiation."[74]

As for the last problem, I., we can consider another hypothetical consequence of the Pauli exclusion principle, and that is degeneracy pressure. It is thought that not only does this principle prevent two electrons from occupying the same state, but that it can lead to a pressure force. This force is thought to be important to the development of white dwarf stars.

What that really tells us is that there is a repulsive force between electrons. Since the Pauli exclusion principle is not a fundamental force, this is a new type of force. It is also important to note that the Pauli exclusion principle cannot be responsible for the force between electrons and protons, as it does not apply between different particles. A complete quantum mechanical wave equation should account for this force as well.

That gives us another new force for our list. They are the:

1. mechanical force,
2. electromagnetic force,
3. weak force,
4. strong force,
5. dark force,
6. high-energy gravity force,
7. repulsive force between electrons and protons,
8. repulsive force between electrons.

The statement that the Schrödinger equation describes electron motion is a lie. From our list we can see that there are numerous gaps in the theory that can only be solved by introducing new physical explanations.

[72] P.F. Browne, "The Cosmological Views of Nernst: An Appraisal." Apeiron, Vol. 2 N. 3, July 1995.
[73] E. Schrödinger, E., "An Undulatory Theory of the Mechanics of Atoms and Molecules." Physical Review. 28 (6): 1049–1070, 1926.
[74] R. Fleming, "Quantum jumps as vacuum fluctuation particle pair interactions analogous to Hawking radiation." gsjournal.net. 29 Aug. 2015.

Lie #58: Physics Explains Photon Production

The 58th of the greatest lies in physics is physics explains photon production. Photon production and absorption are two of the most important interactions in physics. But even though physicists may think they know how and when photons form, there is no accepted physical explanation for it.

Understanding the physical mechanism responsible for photon production is an important problem in physics. It is thought that an electron, in a classically smooth Bohr type orbit, will have to emit photons and radiate energy. There is however no model for photon radiation at the particle level. Schrödinger's wave equation was thought by some to fix this problem, but once again, he did not know enough about how a photon is produced to make a true determination.

In the last chapter we saw that Browne, based on the work of Planck and Nernst, concluded that stable non-radiating orbits could occur if the electron is stabilized by interactions with the zero-point field.[72] Note that quantum electron-positron pairs do rotate at the speed of light without radiating.

Interestingly, we do not actually see particles radiate photons when orbiting or traveling along curved paths when influenced by electric or magnetic fields. This classical assumption appears to be false, based on the absence of experimental data. **The statement that particles in orbits or traveling along a curved path must radiate photons is a lie.** Instead we have a situation where particles are in equilibrium with the aether so they do not need to radiate. The same is true for planets orbiting the Sun. When we do see photon radiation there is a sharp change in direction.

Annihilation-production interactions between electrons and quantum electron-positron pairs give us a physical model allowing us to understand how electrons move, or in particular jump without emitting a photon. When this happens the electron truly is in equilibrium with the zero-point field and does not radiate.

Figure 57-2 in the previous chapter gives us a model for how photons are emitted during orbital transitions. Whenever an electron and positron annihilate while having non-zero energy, they start a new photon. Since a photon is composed of a series of quantum electron-positron pairs, it makes sense that a photon starts out the same way, with the annihilation of an electron-positron pair.

Hypothetically a photon can start with the annihilation of any real particle pair with non-zero energy such as a proton-antiproton pair. Such a proton-antiproton photon may account for certain very high-energy photons.

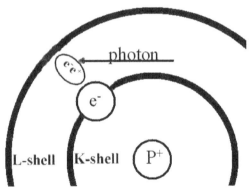

Figure 58-1 A Bohr style hydrogen model showing a quantum electron-positron from a photon about to interact with an orbital electron and cause it to jump to a higher orbit.

The electron orbital transition can be looked at in reverse in order to explain photon absorption. When a photon impinges on an orbital electron, it may be absorbed if it has the proper energy to cause the electron to jump to a higher orbital, or be ejected from orbit completely. Of course, the jump in this case is a quantum electron-positron mediated annihilation production interaction, but in this case, the electron and positron come from the photon instead of the zero-point field. The start of such an interaction is illustrated in figure 58-1.

Black body radiation is another important example. We cannot figure out how the black body radiation of the CMB is emitted, or any other black body for that matter, if we do not understand how black body photons form. Once we understand that quantum particle pairs start the process of black body radiation in a vacuum, it is easy to figure out the rest.

Thermal electrons in the body of matter have energy in relation to the body's temperature. When a quantum electron-positron pair gets too close, the positron annihilates with the electron producing a photon that includes some or all the electron's thermal energy. The once virtual electron becomes free and caries the left-over energy.

It should be very clear that physicists who think the standard model explains how photons are produced and absorbed are wrong. Once we understand that space is filled with quantum particle pair dipoles, and that photons are quantum particle pair dipoles, it is easy to physically explain photon production and absorption.

It is a lie to say that physics explains photon production prior to the existence of the quantum particle pair annihilation-production model.

Lie #59: The Neutral Pion Quark Model

> *If h-bar is fundamental, e will have to be explained in some way in terms of the square root of h-bar, and it seems most unlikely that any fundamental theory can give e in terms of a square root, since square roots do not occur in basic equations.*[75]
>
> Paul Dirac, 1963

The 59th of the greatest lies in physics is the neutral pion quark model. The neutral pion is important to physics, as it is the most elementary of the metastable particles or resonances. They are considered metastable because of their short lifetime, which is on the order of 8.4×10^{-17} seconds. That is still a long time for a particle that is not permanently stable, when compared to other unstable particles which all have even shorter lives.

Neutral pions are really not particles at all, but rather, a resonant state of something else, with that something being another particle or combination of particles. In the 1960s it was still common for physicists to leave open the question of whether a short-lived particle was a particle or just a resonance, but since Nobel Prizes are more readily awarded for discovering particles than resonances, physicists stopped calling them resonances. That is a shame, as we can figure out how to derive 122+ different masses as part of a resonance model, but not from a particle model.

If a particle is a resonance, the first way to determine its composition is to look at what it decays into. The common decay paths for the neutral pion are:

1. two photons,
2. an electron and positron,
3. a photon plus an electron-positron pair,
4. two electron-positron pairs.

So, from the decay paths, the simplest way to model a pion would be to start with the assumption that it is an electron-positron pair in some metastable state. An electron-positron pair can then interact with a quantum electron-positron pair to produce each of the four possible sets of decay products.

But instead, physicists adopted the quark model and within the quark model the neutral pion is equal to something quite different. The right side of that equation is shown in the figure below.

Figure 59-1

$$\frac{u\bar{u} - d\bar{d}}{\sqrt{2}}$$

In the quark model the neutral pion is supposed to be made of a down and anti-down quark pair subtracted from an up and anti-up quark pair divided by the square root of 2. This formula brings up some key questions.

   A. When a quark and its anti particle combine, do they annihilate leaving nothing but energy and momentum?
   B. If there is something left over, beyond energy and momentum when an elementary particle-anti-particle pair annihilate, are they actually elementary?
   C. If one elementary particle can be subtracted from another elementary particle leaving some net difference of something, can they truly be considered elementary particles?
   D. If an elementary particle is divisible by the square root of two, can it truly be elementary?
   E. If a particle equation is divisible by a square root is it irrational?

The answers to the first two questions are yes and no. In essence the formula is contrived to allow for two photons to be the decay product, but this combination of quarks cannot be a particle, even a short-lived one. Of course, fundamental particles cannot be subtracted from one another either. And, as Paul Dirac expressed, if something is divisible by a square root, it is probably not fundamental. The quark model for the neutral pion is truly irrational in both the linguistic as well as the mathematical sense.

Note that up and anti-up quarks orbiting in a metastable quarkonium state have never been observed and neither has a down and anti-down resonance been observed. Since they do not exist, they cannot form other metastable particles such as the neutral pion.

It turns out that Ernest Sternglass discovered an alternative theory in 1961.[76] Sternglass was inspired to search for a composite model for the pion after reading an earlier attempt by Enrico Fermi and Chen-Ning Yang.[77,78] He recognized, with some prodding by Richard Feynman that the neutral pion could be modeled as a relativistic electron-positron pair. In his resonant pion model, an electron-positron pair is rotating at a speed near the speed of light at relativistic velocity. It is this sort of model that allows us to actually simplify particle theory, and explain the 122+ different masses.

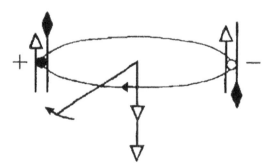

Figure 59-1 A Sternglass style illustration of a neutral pion with an electron-positron pair orbiting each other. Open arrows designate angular momentum and solid arrows designate magnetic moments.

The genius of Sternglass' discovery was in two parts. It was known that as one particle rotates around the other there is a centrifugal force pulling the particles away from each other, and when that force is equal to the electrostatic attractive force pulling the particles together, they have a metastable orbit. The second part was that while there is a low-energy metastable state like this called positronium, there is also a second relativistic metastable state. Actually Feynman may deserve the credit for that idea, based on Sternglass' notes.[77] As Sternglass tells the story he was working out the details on a board in front of Feynman when he derived the pion mass using a simple orbital formula like Bohr used when discovering the quantum states of hydrogen. He later derived the half-life.

Mainstream physicists, notably Alfred Schild, 'disproved' the pion model stating that there was not a relativistic electron-positron pair solution even though quantum fluctuation particle pairs are known to rotate at relativistic velocities. Schild concluded that the relativistic mass of the electron pair must be less than their rest mass.[79] His conclusion is clearly wrong, as the total relativistic mass of a particle moving at relativistic velocities is always greater than the particle's rest mass. He failed at least in part because he did not include the centrifugal force and Thomas precession in his calculation.

Physicists were confused by the Sternglass model and saw the electrons and positron as free particles. They failed to realize the pion in this model is a metastable quantum electron-positron pair. They also continue to complain that a relativistic electron-positron pair cannot exist, totally ignoring the fact that quantum electron-positron quantum fluctuations rotate at speeds up to and including the speed of light all the time. Quantum field theory tells us that relativistic electron-positron pairs do exist.

Another problem is that Sternglass used a Bohr type model which physicists deem to be not sophisticated enough, since it does not address the un-real problem of photon radiation being needed when a particle is in a circular orbit. They said it had to be computed as a wave equation, not realizing that the waves are a property of the zero-point field rather than the particle. While the Sternglass pion model can be improved and made more sophisticated, that does not invalidate the proof of concept. As discussed in the previous chapter, particles following curved paths do not radiate photons, so that part of their complaint is incorrect.

So not only have physicists invented a fictional model of the neutral pion, they have ignored a more successful one. The Sternglass model is consistent with aether being composed of quantum electron-positron pairs, as pions can simply be formed when a quantum electron-positron pair is given enough energy to take on the metastable neutral pion resonant state.

In the end, we simply need this model, in order to be able to model all the particles and resonances. We ultimately must be able to derive 122+ different masses from fundamental principles.

The neutral pion quark model is truly one of the greatest lies of physics

[75] P.A.M. Dirac, "The Evolution of the Physicist's Picture of Nature." Scientific American, May 1963, 208(5), 45-53.
[76] E. J. Sternglass, "Relativistic electron-pair systems and the structure of the neutral meson." Phys. Rev., 123, 391, 1961.
[77] E. J. Sternglass, From particle notebooks in the Ernest J. Sternglass collection of the Kroch Library at Cornell University.
[78] E. Fermi C.N. Yang, "Are Mesons Elementary Particles?" Phys. Rev. 76, 1739, 15 Dec, 1949.
[79] A. Schild, "Electromagnetic two-body problem." Phys. Rev., 131, 2762 1963.

Lie #60: The Other Irrational Meson Models

The 60th of the greatest lies in physics is the other irrational meson models. Mesons are the intermediate particles or resonances with masses that fall between the electron and the proton. Like the neutral pion described in lie #59, there are 6 other mesons that are similarly irrational. Within the scope of the quark model, eta mesons are supposed to be comprised of 8 quarks divided by the square root of 6 with the mathematical relationship shown in figure 60-1.

Figure 60-1

$$\frac{u\bar{u} + d\bar{d} - 2s\bar{s}}{\sqrt{6}}$$

Eta prime mesons are supposed to be composed of 6 quarks divided by the square root of 3 with the mathematical relationship shown in figure 60-2.

Figure 60-2

$$\frac{u\bar{u} + d\bar{d} + s\bar{s}}{\sqrt{3}}$$

K-short mesons are supposed to be comprised of 2 complementary quark pairs subtracted from each other and divided by the square root of 2 with the mathematical relationship shown in figure 60-3.

Figure 60-3

$$\frac{d\bar{s} - s\bar{d}}{\sqrt{2}}$$

K-long mesons are supposed to be comprised of 2 complementary particle pairs, but this time added to each other and then divided by the square root of 2 with the mathematical relationship shown in figure 60-4. There is no explanation for how the K-short and K-long have nearly identical masses when one subtracts the masses and the other adds them.

Figure 60-4

$$\frac{d\bar{s} + s\bar{d}}{\sqrt{2}}$$

Neutral rho mesons have the same components as the neutral pion but are supposed to be different from the pion, including having more mass and a shorter half-life. See figure 60-5.

Figure 60-5

$$\frac{u\bar{u} - d\bar{d}}{\sqrt{2}}$$

Omega mesons are like the neutral rho or pion except the quark pairs are somehow added instead of subtracted as shown in figure 60-6. Again the neutral rho and omega have very similar masses while one subtracts the masses while the other adds them..

Figure 60-6

$$\frac{u\bar{u} + d\bar{d}}{\sqrt{2}}$$

All of these mathematically irrational meson models are also irrational in the literary sense. As with the neutral pion, it is not possible to produce even a metastable particle from particle-antiparticle pairs that have annihilated. Additionally, these pairs cannot be physically subtracted from one each other and quarks, if they are both physically real and elementary, cannot be divisible by a square root of an integer.

The K-short and K-long add another dimension of irrationality. A down quark and anti-strange quark is equivalent to a kaon meson. A strange quark and an anti-down quark form an anti-kaon. So if we add a kaon and anti-kaon they should annihilate and yield nothing, expect perhaps excess energy and momentum. Kaons are also not divisible by the square root of 2.

Within the scope of the quark model, and including the neutral pion, there are 7 mesons modeled in this irrational way. These other meson models are collectively one of the greatest lies in physics. They are not physical descriptions of these known resonances.

The question follows that if these mesons are not made of quarks, then what are they? It turns out that the masses of the particles, both mesons and the heavier hadrons, are related to each other by values which are close to multiples of the inverse of the fine structure constant, ~137. With some work it is therefore possible to model each meson and baryon—other than the electron, proton, and neutron—using one or more relativistic electron-positron pairs plus additional electrons and nucleons where necessary. This is accomplished in my book *Goodbye Quarks: The Onium Theory*. The book includes models for the resonances listed in this chapter showing that the irrational quark meson models are a lie.

Lie #61: Quarks

> *Ne'eman:* Can I ask you something that Feynman asked me some time ago: Would you be ready to drop your theory if the 720 meson turns out to have spin 0 and not spin 2, or if there is another nucleon state there?
>
> *Sternglass:* A theory of this kind either has to fit every particle, or it is not good at all.
>
> *Ne'eman:* This is certainly a satisfying answer.[80]
> Yuval Ne'eman & Ernest Sternglass, 1965

The 61st of the greatest lies in physics is quarks. The above exchange between Yuval Ne'eman and Ernest Sternglass occurred during the Second Topical Conference on Resonant Particles at Ohio University in June 1965. The quark model had come out the previous year, and Ne'eman along with Murray Gell-Mann had independently developed the numerological symmetry theory that led up to the quark theory. As such Ne'eman was heavily vested in the quark theory and set about to attacking Sternglass' theory at the conference. As fate would have it, it is the quark model that falls to properly account for the properties of each of the mesons. "A theory of this kind (i.e. the quark model) either has to fit every particle, or it is not good at all." To be fair, the Sternglass model fails the test too as only his neutral pion, charged pion, and muon models are valid.

The most fundamental meson is the neutral pion. So in order to have a valid physical model for the mesons, and all particles/resonances in general, we must start with a valid model for the neutral pion. The system for modeling the mesons begins with the neutral pion and all the other meson models must be based on that.

If your pion model is a lie, which the quark pion model is as explained in Lie #59, then your entire particle model is a lie. Of course, the 6 other irrational meson models discussed in Lie #60 only serve to reinforce the point that the quark model fails the Feynman test.

Some of you may be thinking, "quarks have been experimentally verified to exist," but that is not true. The five lowest mass quarks have never been seen in the laboratory. Their existence is only inferred from experiments if they first assume the quark model is valid. It is the fallacy of wishful thinking at work rather than scientific evidence.

The quark model is a lie. It fails the Feynman test by failing to fit 7 very important mesons. My book *Goodbye Quarks: The Onium Theory* goes into much greater detail on the failings of the quark model. The book also shows how all resonances can be modeled using an onium theory where all short-lived resonances are compounds in relativistic onium states similar to relativistic positronium. Since the quantum field of space contains numerous quantum electron-positron and proton-antiproton pairs, it makes sense to build a particle theory with those as the basic units for all the resonances.

[80] E. J. Sternglass, "New evidence for a molecular structure of meson and baryon resonance states." Proceedings 2nd topical conference on Resonant Particles, Ohio University, Athens, Ohio, June 10-12, [B.A. MUNIR] pp. 55, 1965.

Lie #62: Muons are Elementary Particles

The 62nd of the greatest lies in physics is muons are elementary particles. In the standard model of physics there are three tiers of leptons including three types of electrons: the electron, the mu particle or muon, and the tau particle. Each is treated as an elementary particle, rather than a composite particle.

Early on a mathematical relationship was noted that was good evidence that the muon is a composite particle. While the three pion masses are close in magnitude to the inverse of the fine structure constant, ~137, the muon mass of 105.65 Mev/c² (million electron volts divided by the speed of light squared) is about ¾ths of that. This indicates that there is a simple physical relationship between those four particles. Even the tau has a mass that is almost exactly 13 x 137 MeV/c².

Because of this simple mathematical relationship, it is relatively easy to come up with formulas that are related to composite particle structures or other types of structures, and tweak them to approximate the known masses of the pions and muon. Consequently, numerous scientists have done just that.

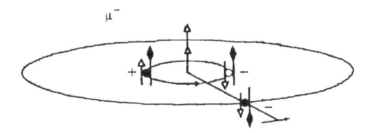

Figure 62-1 A Sternglass style illustration of a muon with an electron-positron pair in the center and an electron in orbit around it. Open arrows designate angular momentum and solid arrows designate magnetic moments.

Ernest Sternglass proposed one of the better models for the muon more than 50 years ago. The Sternglass model is essentially a relativistic electron-positron pair orbiting a non-relativistic electron, although he drew it centered on the electron-positron pair.[81] The Sternglass muon model, is shown in figure 62-1. With the vacuum of space being filled with quantum electron-positron pairs, the Sternglass model is the simplest, and best fitting of the various proposed muon models. His theory shows that it is possible to model all the resonances in a similar manner.

It has been obvious that the muon is a resonant composite particle rather than an elementary particle due to the mathematical relationship and the ease with which it can be physically modeled. Note that Sternglass mentions in his notes that he made several mistakes in his papers from the 1960s, so it is still necessary for someone to pick up his work and fix his mistakes, while otherwise simplifying and improving his model and overcoming certain objections to it.

Physicists have been lying for 50 years about the muon being an elementary particle. Somehow, rather than follow the obvious path of treating the muon as a composite particle, physicists chose to designate the muon as an elementary particle.

---

[81] Sternglass, E. J., "Electron-positron model for the charged mesons and pion resonances." Il Nuovo Cimento, 1 Gennaio, Volume 35, Issue 1, pp 227-260, 1965.

Lie #63: Tau Particles are Elementary

The 63rd of the greatest lies in physics is tau particles are elementary. The tau is said to be elementary and the third-tier electron within the standard model. As discussed in Lie #62 there was a pretty good model for the muon over 50 years ago which the physics community has generally ignored. I attempted to model the tau as a higher-energy relativistic muon some years ago but since then have determined that early model is invalid.[82,83]

The tau decay modes give us a picture of the tau that is much more complicated than a simple excited muon. Some of the most common decay modes for the tau are shown in Equations 63-1 in descending order of probability.[84] While decay modes (b) and (c) are consistent with a tau being an excited muon, the others are not as it can decay to one to four pions.

Equations 63-1

a. $\tau^- \to \pi^0 + \pi^-$ or $\tau^+ \to \pi^0 + \pi^+$ (25.49%)
b. $\tau^- \to e^-$ or $\tau^+ \to e^+$ (17.82%)
c. $\tau^- \to \mu^-$ or $\tau^+ \to \mu^+$ (17.39%)
d. $\tau^- \to \pi^-$ or $\tau^+ \to \pi^+$ (10.82%)
e. $\tau^- \to 2\pi^0 + \pi^-$ or $\tau^+ \to 2\pi^0 + \pi^+$ (9.26%)
f. $\tau^- \to \pi^+ + 2\pi^-$ or $\tau^+ \to \pi^- + 2\pi^+$ (8.99%)
g. $\tau^- \to \pi^+ + 2\pi^- + \pi^0$ or $\tau^+ \to \pi^- + 2\pi^+ + \pi^0$ (2.74%)
h. $\tau^- \to \omega + \pi^-$ or $\tau^+ \to \omega + \pi^+$ (1.95%)
i. $\tau^- \to 3\pi^0 + \pi^-$ or $\tau^+ \to 3\pi^0 + \pi^+$ (1.04%)

The tau has eight decay modes to two kaons, each of which contains two pions in onium theory. Kaons are pionium. In total, the tau has 30 decay modes to four pions indicating that it must contain four pions rather than an excited single muon or pion. In the onium theory, as discussed in the referenced paper and my book *Goodbye Quarks: The Onium Theory*, it is shown that tau particles contain two kaons in relativistic orbit centered around an electron or positron. One possible combination is shown in figure 63-1.[84] It is also possible for the tau to contain a positive and negative kaon pair.

Figure 63-1 A negatively charged tau with a positron orbited by two relativistic kaons.

In the onium theory the relativistic mass from the two-kaon orbit is a factor of m_e/α per particle where m_e is the electron mass and α is the fine structure constant. Each kaon contains two charged pions and an electron or positron for a total of seven particles per kaon or 14 in a two-kaon orbit which produces 980.35 MeV/c² in relativistic mass. There are two types of two-pion resonances where one or both pions is in relativistic orbit. The neutral kaon mass is 497.614 MeV/c² when both are relativistic, and ~100-105 MeV/c² less when only one is relativistic, in what I call the K_D state.

In the tau the two kaons are in the K_D state such that two times the K_D mass, plus m_e and 980.35 MeV/c² in relativistic mass equals the tau mass of 1776.84 MeV/c². In the tau each K_D has ~398 MeV/c² in mass. The relativistic K-K_D and K-K resonances are D mesons with masses of 1969.61 ($D^{\pm}$), 1964.84 (D^0), and 1968.30 ($D_s^{\pm}$) MeV/c². The nested K_D-K_D and K-K_D resonances are the of 782.65 MeV/c² omega, 891.61 MeV/c² K*(892)$^{\pm}$, and 895.81 MeV/c² K*(892)^0. The phi and f$_0$(980) mesons are nested K-K resonances.

Physicists are lying when they say that tau particles are elementary. Out of the electron, mu, and tau only the electron is elementary.

---

[82] R. Fleming, "A tau particle model based on the Sternglass theory." https://www.researchgate.net.
[83] R. Fleming, "An onium model of particles with only electrons and protons." gsjournal.net, 19 Nov. 2019.
[84] M. Tanabashi et al. (Particle Data Group) (2018) Phys. Rev. D 98, 030001.

Lie #64: Electron Neutrinos are Elementary

> *As the bearer of these lines, to whom I graciously ask you to listen, will explain to you in more detail, because of the "wrong" statistics of the N- and Li-6 nuclei and the continuous beta spectrum, I have hit upon a desperate remedy to save the "exchange theorem" (1) of statistics and the law of conservation of energy. Namely, the possibility that in the nuclei there could exist electrically neutral particles, which I will call neutrons, that have spin 1/2 and obey the exclusion principle and that further differ from light quanta in that they do not travel with the velocity of light. The mass of the neutrons should be of the same order of magnitude as the electron mass and in any event not larger than 0.01 proton mass. The continuous beta spectrum would then make sense with the assumption that in beta decay, in addition to the electron, a neutron is emitted such that the sum of the energies of neutron and electron is constant.*[85]
>
> Wolfgang Pauli, 1930

The 64th of the greatest lies in physics is electron neutrinos are elementary. When scientists were studying radioactive decay, they found that it violates the principle of conservation of energy as they understood it at the time. The detectable decay products have a range of energies. For example, when a neutron decays it releases the ~780 keV needed to produce it, but the median electron energy is less than 400 keV. In their minds the decay had to have a single fixed maximum energy (e.g. ~780 keV), and so energy was being lost somehow when a lower energy particle was emitted. There was no detectable particle that carries away that energy so they found it necessary to invent one.

The alternative to inventing a particle, they thought, was to accept that the principle of conservation of energy is not true in every situation. Since physicists were unaware of a physical mechanism to explain beta decay, and were not ready to give up on conservation of energy, inventing a new particle seemed to be the way to go. For those who are not familiar, note that beta particles are the same as electrons, and beta decay encompasses decay interactions involving both electrons and positrons.

This invented particle had to be electrically neutral, very small, and with little or no mass, or else we would see it directly. Wolfgang Pauli was the first to propose the existence of such a particle, which he called neutrons in the letter quoted at the beginning of this chapter. Enrico Fermi further developed Pauli's concept, and came up with the neutrino name.[86]

Over time, with the discovery of excited electron states that are called the mu and tau electrons, it was thought there must also be mu and tau neutrinos, and hence we have the three neutrinos of the standard particle model. Additionally, there are antineutrinos, such that lepton conservation can be preserved as an anti-lepton cancels out a lepton in that rule-based formula. Since under this theory neutrinos are being emitted all the time, there is thought to be a field of neutrinos throughout the universe, holding an abundance of nearly undetectable energy, and perhaps even mass.

Neutrino theory, however, does not give us a physical model for how beta decay actually happens, explain how there is an energy distribution, nor how that distribution physically occurs. In the current theory the energy distribution just somehow magically happens and the neutrino takes up the slop.

To actually explain beta decay and its energy distribution, this new particle, or whatever, must come at the front end of the interaction and must contribute energy to the interaction. Then the emitted particle, typically an electron or positron, has the leftover energy.

Neutrinos are on the wrong side of the equation to be the actual physical cause of beta decay and its energy distribution. **Standard model physicists lie if they say they understand the cause of beta decay.**

Standard model physicists like to pretend that they have addressed this fundamental problem with weak interaction theory, but they have not. **They lie if they say they know the cause of the energy distribution and lie if they say a neutrino is responsible for the energy distribution.**

So what really causes beta decay and the energy distribution? What is required is not far from what Pauli proposed. We need a bunch of 'little neutral things' that have a broad range of energies. These 'little neutral things' must interact with nuclei subject to decay in accordance with a probability cross-section. These 'little neutral things' must be ever present throughout the universe. They must form a field throughout space.

This field is not neutrinos, as there is no mechanism for individual free neutrinos to be produced throughout space. Neutrinos also do not have an obvious physical mechanism for say, grabbing the electron-like part of a neutron and freeing the electron.

Fortunately, there is a field that meets our requirements, the zero-point field, and it has dipoles that do meet the 'little neutral things' requirement. Quantum fluctuations are electrically neutral when taken as a whole; they have no mass, and are not directly detectable. They also have a continuum of energies.

There is also an obvious interaction mechanism, which is much like the mechanism behind quantum jumps shown in chapters 57 and 58. A free electron annihilates with a quantum positron and the once quantum electron becomes free. By extension, the electron-like part of a neutron annihilates with a positron turning the neutron into a proton and leaving the once quantum electron free. In this manner, an electron is seen as jumping across the electron-proton ~780 keV potential barrier, and escaping its confinement with the proton.

This process is readily reversed so that an electron can jump through the potential barrier surrounding a proton to produce a neutron. In this case the interaction with the quantum electron-positron pair makes it appear that the electron has tunneled through the potential barrier. This is a specific case of a broader class of quantum fluctuation particle pair mediated annihilation-production events called quantum tunneling.

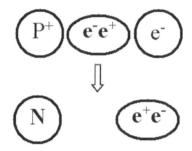

Figure 64-1 A neutron is formed by an electron-positron annihilation-production interaction

Annihilation-production interactions involving quantum fluctuation particle pairs—electron-positron as well as proton-antiproton—account for all beta decay interactions. For more details please refer to some of my papers.[87,88,89]

Interestingly, the missing energy comes from the zero-point field. In the case of neutron production, zero-point energy is incorporated into the neutron, and when a neutron decays, energy is released back to the zero-point field. But, since the neutron's energy is displaced zero-point energy to begin with, there is no conservation of energy violation. Zero-point energy is energy. Annihilation-production events allow zero-point energy to also balance momentum as it does in atomic orbits.

Louis De Broglie hypothesized that photons are composed of electron-positron pairs and later neutrino pairs with there being a new neutrino pair each half wavelength. Since a photon consists of a quantum electron-positron pair each half wavelength, we can instead conclude that a neutrino looks like a quantum electron-positron pair in some respects. This brings us full circle back to weak interactions being mediated by quantum electron-positron pairs.

But you may ask, what about the neutrino detection experiments? What are they detecting? The answer is, they are mostly detecting the zero-point field. After all, there does not need to be both a zero-point field and a neutrino field when the zero-point field alone can account for all the physical phenomena.

That said, there is a hypothetical model for neutrinos that allows them to carry linear and angular momentum, while at the same time being part of the zero-point field. In a twist on the De Broglie hypothesis, a neutrino, like a photon, is composed of a series of rotating quantum electron-positron pairs, but instead of counter-rotating, they rotate in the same direction conserving angular momentum. A composite neutrino model allows for neutrinos to be emitted and detected and makes antineutrinos the same as neutrinos.

In any case, stating that electron neutrinos are elementary is a lie. Electron neutrinos are nothing more than quantum fluctuation dipoles or combinations of quantum fluctuation dipoles, and it is the quantum fluctuations that are elementary.

[85] W. Pauli, "Open letter to the group of radioactive people at the Gauverein meeting in Tübingen." 4 Dec. 1930. (Translation: Kurt Riesselmann)
[86] E. Fermi, Z. Physik 88 161, 1934.
[87] R. Fleming, "Beta Decay as a Virtual Particle Interaction Analogous to Hawking Radiation." researchgate.net.
[88] R. Fleming, "Neutrinos as vacuum fluctuation particle pairs." researchgate.net.
[89] R. Fleming "Quantum jumps as vacuum fluctuation particle pair interactions analogous to Hawking radiation." gsjournal.net. 29 Aug. 2015.

Lie #65: Mu & Tau Neutrinos are Elementary

The 65th of the greatest lies in physics is mu and tau neutrinos are elementary. Since lie #64 is electron neutrinos it is an obvious consequence that calling mu and tau neutrinos elementary is also a lie. But there are important differences. As mentioned in the last chapter, neutrinos are thought to exist in three forms that correspond to the form of electron with which they interact. The three electron types are the electron, mu, and tau. Consequently, there is thought to be an electron neutrino, a mu neutrino, and a tau neutrino that interact with their respective electron type.

For each of these neutrino types there is a complementary anti-neutrino; however, the anti neutrino may be the same as a neutrino. This may be similar to the way an anti-photon is the same as a photon. In this respect, and as suggested in the last chapter, the similarities between photons and neutrinos are much greater than proposed by the standard model.

Mu neutrinos were discovered in 1962 by a group of researchers who went on to win the 1988 Nobel Prize for their discovery.[90] What they discovered was that when what was thought to be neutrinos emitted during pion decay interacted with matter, they produced muons rather than electrons. A normal electron neutrino would be expected to produce an electron when it interacts with matter. Note that the muon will then decay to an electron anyway.

From the previous chapter and the related papers, it is easy to see how quantum electron-positron pairs fill the role of neutrinos for modeling beta decay processes. A regular electron neutrino does not have to come from the decay of another particle as it can be any quantum electron-positron pair in the aether that falls within the proper energy range.

What is not immediately clear is how a decaying pion leads to muon production and only muon production. It is easier to see what is going on when we look at the Sternglass models for the pion and muon shown in the figures below.

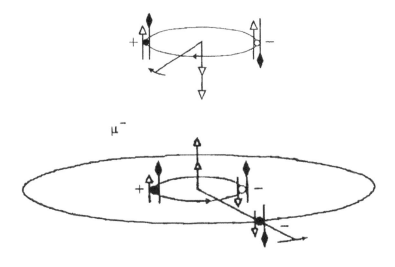

Figure 65-1 Sternglass style illustrations of a neutral pion (top) and a negative muon (bottom). The open arrows indicate angular momentum and the closed arrows indicate magnetic moment.

Of course, the first thing that jumps out is that there is a rapidly rotating electron-positron pair in both particles. That means something about pion decay causes a pion-like electron-positron pair to be reestablished forming a muon. This looks like a conservation of angular momentum issue.

We also need to consider what a pion decays into, which is two photons, an electron and positron, a photon plus an electron-positron pair, or two electron-positron pairs. Charged pions will decay to a muon as an intermediate step to finally decaying to an electron. In the Sternglass model, a charged pion is essentially an excited muon. In the standard model there will be neutrinos or antineutrinos emitted during each step in the decay process.

The question is, what happens to the pion's angular momentum? This angular momentum is something that should be conserved. Photons or individual electrons or positrons cannot carry it away. Free electron-positron pairs may carry a much smaller amount of angular momentum as positronium. None of the directly detectable decay products appears to be the answer.

The only option left is that the pion's angular momentum must be carried away by the mu neutrino, whatever it is. But, in order to carry that much momentum, the mu neutrino must itself be composed of a series of relativistically rotating quantum electron-positron pairs. It ends up being like a photon, but instead of having each successive pair counter-rotate they must rotate in the same direction, thereby conserving angular momentum. This is consistent with the model mentioned in the last chapter.

The tau particle carries even more angular momentum than a mu particle due to its higher mass. Consequently, a tau neutrino must carry more angular momentum as well. This large amount of angular momentum also gives us an explanation how mu and tau neutrinos do not interact with ordinary matter very often, as there is generally no way for the angular momentum to be absorbed by most particles, atoms, or molecules.

Quantum electron-positron pairs, in the guise of electron neutrinos, do not interact with ordinary stable matter. They interact with atoms that are subject to beta decay processes. Note that, because quantum fluctuations are everywhere, we will often see so-called electron neutrinos while expecting to see mu or tau neutrinos.

Based on the onium theory the mu and tau are both mesons. So in that case the logic that the mu and tau need their own special neutrinos would mean that all mesons and baryons need their own special neutrino. That is obviously not the case.

As an additional note, neutrinos could hypothetically be composed of other real quantum particle pairs when such pairs are part of an interaction. Although based on the decay products of all the resonances, they mostly appear to be made with electron-positron pairs.

Like the electron neutrino, stating that mu neutrinos and tau neutrinos are elementary is a lie. All neutrinos can be accounted for as composite particles consisting one quantum particle pair or of a series of quantum particle pairs.

[90] G. Danby, J. M. Gaillard, K. Goulianos; L. M. Lederman; N. B. Mistry; M. Schwartz; J. Steinberger, "Observation of high-energy neutrino reactions and the existence of two kinds of neutrinos." Physical Review Letters. 9: 36, 1962.

Lie #66 Neutrinos have Mass

The 66th of the greatest lies in physics is neutrinos have mass. In physicists' desperation to save the doomed gravitational theories of Newton and Einstein, they have jumped the gun in deciding that neutrinos have mass, as neutrino mass could be the missing dark matter. As has been discussed in previous lies, the gravitational theories are beyond saving and there is no reason to be searching for dark matter or missing mass. There were one or more forces missing from those models all along.

Physicists have decided that neutrinos have mass because they oscillate between the three types, so a mu neutrino may later appear as an electron neutrino. The way these are tied together is not explained in any physically logical manner, since they are not based on a physical explanation. The neutrino mass is said to be low, but there is no mechanism for how a neutrino could displace very long-wavelength, low-energy quantum fluctuations.

If we consider beta decay, which is best explained as a quantum fluctuation interaction, those 'neutrinos' certainly have no mass. They are quantum fluctuations and quantum fluctuations by themselves have no mass.

A neutrino that is composed of a series of quantum fluctuations does not have mass either. Photons do not have mass, and neutrinos that are similar to photons except for their angular momentum, would not have mass. The mass of this type of neutrino does not equate to its energy, just as the mass of a photon does not equate to a photon's energy. The oscillations that have been observed are best explained as transfers of angular momentum between these photon-like neutrinos.

The idea that neutrinos have mass is a lie. There is no field of neutrinos throughout space accounting for missing mass, which is not really missing.

Lie #67: W & Z Bosons Mediate Weak Interactions

The 67th of the greatest lies in physics is W and Z bosons mediate weak interactions. The hypothesis that W and Z bosons mediate weak interactions has numerous faults. This hypothesis is so bad that it truly makes one wonder about the intellectual capacity of those who believe in it. As we saw in Lie #18 there is a long list of problems with the gauge boson model of force interactions. The W and Z boson model adds even more problems.

Many of the additional problems with the W and Z relate to their mass-energy of 80 GeV (giga-electron-volts) for the W and 91 GeV for the Z. Both are more than 85 times more massive than a proton. But ultimately the theory fails because there is a much simpler theory described briefly in chapter 64, which explains what happens in a simple physical model, and without the numerous flaws of the W/Z theory.

Massless photons are bad enough as gauge bosons, but when you have a gauge boson with mass things get much worse. You cannot get a stable or meta-stable W or Z particle from a much less massive proton or electron without violating the principle of conservation of energy. The exception would be if it were so short lived that it was a quantum fluctuation existing as a matter-antimatter particle pair that does not exceed the Plank energy.

The next problem, relating to both mass and energy conservation, is path length. Those who favor the hypothetical W/Z theory say that limited path length is a part of the theory that matches weak interaction observations. The problem is that the allowable path length that W and Z particles could have without violating the principle of conservation of energy is far too short. This is a serious problem, not a positive feature.

These hypothetical gauge bosons cannot exist long enough to escape a proton or neutron, as they cannot exist longer than 1.3×10^{-26} seconds without violating the principle of conservation of energy. During that time a W particle can at most travel 3.9×10^{-18} meters, and the Z less than that. In comparison, the proton radius is 0.8775×10^{-15} meters and the neutron radius is similar. In order for a neutron to decay into a proton and electron, the electron must get outside the proton's ~780 keV potential barrier, which is at least as large as the radius of a proton. The W/Z model does not work.

Additionally, the W/Z hypothesis violates experimental measurements of those particles' mean lifetimes, as that also limits their travel distance. During a 3×10^{-25} second mean lifetime, a W can only travel 9×10^{-17} meters if it is moving at the speed of light, which is about a tenth of the radius of a proton. A W particle would have to violate the principle of conservation of energy, and/or the speed of light limit to allow an electron to escape a proton.

Even if the W or Z existed as a quantum matter-antimatter particle pair, the creation of an electron as part of the decay would once again violate the principle of conservation of energy. The theory, however, does not even contemplate the particle pair option, so it violates conservation of energy no matter how you look at it.

Fortunately, there is a much simpler mechanism to explain weak interactions. Annihilation-production interactions mediated by quantum particle pairs account for all beta decay interactions. The W and Z are simply not needed to explain weak interactions.

The idea that W and Z bosons mediate weak interactions is a lie. There is no way for this model to comply with the principle of conservation of energy and it is not even necessary.

Lie #68: W & Z Particles

The 68th of the greatest lies in physics is W and Z particles. Based on the previous lie it should be crystal clear that W and Z particles do not mediate weak interactions, so it should be no surprise that the true nature of W and Z particles must be questioned.

Quantum fluctuation particle pairs, primarily quantum electron-positron pairs, mediate all weak interactions. Even if the W and Z interaction theory were believed, it violates the principle of conservation of energy. The smart particle gauge boson theory is also a lie as discussed in lie #18, even without these additional difficulties.

I bet some of you are thinking 'but there is experimental evidence for W and Z particles. They have been proven to exist.' That, however, is not what the physical evidence implies at all. The experiments detected short-lived resonances with a $\sim 3 \times 10^{-25}$ second half-life. There is no direct physical evidence as to what these resonances are, or what they do. The W and Z are not even detected directly, but are identified based on their decay products. To even call them a particle is unscientific. They barely meet the requirements of being statistically significant measurements of resonances.

There is no direct physical evidence that these resonances participate in the reactions they are said to participate in. Scientists simply chose to match up these resonances with an existing theory, simply because they wanted to believe it was true, and they wanted a shot at winning a Nobel Prize. Saying that these resonances participate in weak interactions is unscientific dogma.

As of now there is no commonly accepted physical model for these high-energy resonances. But if we consider the Feynman-Sternglass pion model with a relativistic proton-antiproton pair instead of an electron-positron pair it would have a mass of around 247 GeV/c^2. Under the onium theory there should be resonances with relativistic protons where the proton adds $1/\alpha$, $3/4\alpha$ or $1/2\alpha$ times the proton mass adding approximately 128, 96, or 64 GeV/c² to a resonance. Consequently, relativistic proton resonances likely account for the W, Z and Higgs bosons.[91] And since the mass of the top quark equals the sum of the masses of the W and Z, relativistic protons can account for the top quark mass as well.

W and Z particles are a lie. **W and Z particles are not elementary particles.** They are nothing more than the resonant state of a combination of two or more particles containing a relativistic proton.

[91] R. Fleming "The W, Z, Higgs, and top as relativistic protonium." gsjournal.net. 19 Nov. 2019.

Lie #69: The Strong Nuclear Force is due to Gluons

The 69th of the greatest lies in physics is the strong nuclear force is due to gluons. The strong nuclear force is the force that binds protons and neutrons together in an atomic nucleus. Without the strong nuclear force, the electrostatic repulsion between protons would cause atomic nuclei to split up when there is more than one proton in the nucleus.

The nuclear force has to be strong enough to hold protons near each other even though they are repelled due to their positive electric charge. This force is up to 100 times stronger than Coulomb repulsion, but when they get too close, on the order of 0.7×10^{-15} meters, there is a repulsive force. Even so, two protons will not remain attracted to each other without there also being at least one neutron in the nucleus. Most of the stable atomic nuclei have at least as many neutrons as protons in them.

In the quark theory, gluons, as part of the closely related strong force, are said to hold quarks together to form particles, most importantly protons and neutrons. A residual part of this gluon-mediated force is then supposed to account for the nuclear forces that hold protons and neutrons together within the nucleus.

Prior to the quark theory, mesons were the gauge bosons of strong force theory. But now, under the standard model of physics, gluons are said to be the gauge bosons of the strong force. Whether we are talking about the gluon model or the meson model for strong forces, gauge bosons are in general a lie for the reasons discussed in Lie #18. **So, if someone goes 'old-school' and says that mesons mediate the strong nuclear force, they are still lying.**

More importantly there is a far more elegant alternative explanation for the strong force, an explanation that also unifies the strong force with electromagnetic theory. This is a force that always exists between objects in close proximity to each other, the Casimir force.

While the original Casimir theory predicted an effect between two plates, the first convincing experimental confirmation confirmed the force existed between a sphere and a plate. The Casimir effect also occurs between two spheres, and protons have been found to be spherical or quasi-spherical objects in scattering experiments. Neutrons have a spherical shape as well, of about the same size. Given this, there must be Casimir forces between protons, between neutrons, and between protons and neutrons.

It turns out that if you calculate the Casimir force between two spheres the size of two protons, it is strong enough to account for the nuclear force, while also overcoming electrostatic repulsion. A simple computation that addresses part of the force shows the attractive Casimir force between two spheres to be at least 81 times the electrostatic repulsion at 0.5 fm. For more information read my paper "the Nuclear Force Computed as the Casimir Effect Between Spheres."[92]

Note that the paper only describes one component to the total Casimir force as it only includes pressure differentials at the nearest boundaries between the protons. The paper does not consider that the spherically shaped particles are themselves Casimir cavities, and there is additional pressure coming from the outer surfaces of the particles. When all contributions to the Casimir force are considered the forces should total approximately 100 times the Coulomb repulsion at 0.7 fm.

The repulsive force that takes over when the particles get very close is said to be part of the strong force, but the two forces, acting in opposite directions, are best treated as two separate forces. Some physicists claim that the repulsive force is due to Pauli's exclusion principle. Pauli's exclusion principle is not a fundamental force and does not explain the repulsive force between protons and neutrons. **The Pauli's exclusion principle explanation of the repulsive force between nucleons is yet another lie.**

So far in this book we have identified repulsive forces between electrons, protons, and neutrons, and between protons and electrons and neutrons and protons. This is collectively an entirely different force from the strong force. **To say that the repulsive force between protons is part of the strong force is another lie.**

So now we must once again amend our list of the fundamental forces. They are the:

1. mechanical force,
2. electromagnetic force,
3. weak force,
4. strong force,
5. dark force,
6. high-energy gravity force,
7. repulsive force between electrons and protons,
8. repulsive force between electrons,
9. repulsive force between protons,
10. repulsive force between protons and neutrons,
11. repulsive force between neutrons.

The idea that the strong nuclear force is due to gluons is a lie, as is the older idea about it being due to mesons. Strong nuclear forces are Casimir forces, and equally important, strong nuclear forces are not a separate force, but are part of the electromagnetic force.

[91] R. Fleming, "The Nuclear Force Computed as the Casimir Effect Between Spheres." gsjournal.net. 24 Jun. 2019.

Lie #70: Gluons

The 70th of the greatest lies in physics is gluons. Once physicists had convinced themselves that photons were exchange particles, or gauge bosons, which carried the electromagnetic forces, they liked the idea so much they took to inventing new gauge bosons for all the known forces. Originally Yukawa came up with the Nobel Prize winning idea that mesons, or more specifically the newly discovered pions, were the gauge bosons of the strong nuclear force.[92]

Once the quark theory took hold, physicists felt the need to invent a new gauge boson that acts as a force exchange particle between quarks. Unfortunately, they never thought very critically about the logic, or lack thereof, behind the exchange particle hypothesis. Please review Lie #18 if needed.

Now, since it can easily be shown that the Casimir force accounts for the strong force, there is no need for a gauge boson to be associated with the strong force. The Casimir force is due to van der Waals forces between quantum fluctuation dipoles that are known to exist regardless of any other hypothesis.

Interestingly, quantum electron-positron and proton-antiproton pairs do cause neutrons to convert to protons and protons to convert to neutrons within the nucleus. The exchange of electrons and protons is mediated in the same way that we see with neutron production and decay, and due to the close spacing within the nucleus, both steps happen rather instantaneously.

Gluons are a lie. **The older idea that mesons are gauge bosons is also a lie.** There is no need for force exchange particles to explain the strong force.

[92] H. Yukawa, "On the Interaction of Elementary Particles," PTP, 17, 48, 1935.

Lie #71: Electrons do Not Feel the Strong Force

The 71st of the greatest lies in physics is electrons do not feel the strong force. Note that this is not to say that electrons interact with protons and neutrons by the strong nuclear force, because they do not, for reasons that will become clear. No, the point is that there is an equivalent strong force that attracts electrons to each other. There is strong evidence for this electron binding force, and yet for some reason it is not included in the standard model of mainstream physics.

What is the evidence you ask? Consider lightning as an example. Lightning looks like streaks of light in the sky to our naked eye. In reality those streaks are made of electrons and those electrons travel in a packet. As electrostatic charge builds up, a group of electrons collect whether in a cloud or on Earth, and when the difference in electric potential, the voltage difference, is great enough, an arc forms and a packet of these electrons jump across the potential gap.

Note that this happens anytime there is an arc, even when you shock yourself due to build-up of static electricity from rubbing against something. The electron packet moves so fast that our eyes see a line instead of a dot or ball, with the exception of rare slow-moving phenomena such as ball lightning.

The question then is how does this packet form? If we consider only forces in the standard model, we would have a packet filled with negatively charged electrons with every electron repelling every other electron. If the standard model described all the forces, those electron packets would blow up. There would be no packet of electrons, no arcs, and no lightning.

For our common everyday observations to be true, there must be an additional force that causes electrons to be attracted to each other. While a few researchers have looked into this problem, there is presently no widely accepted theory to explain it. Interestingly enough, the same force as the strong force theory presented in Lie #69 works.

The force that overcomes the electrostatic repulsive force between electrons is the Casimir force. If electrons are close enough together, the Casimir force binds them together, overcoming electrostatic repulsion, in the same way that protons are bound together. The reason that electrons and protons are not pushed together by the Casimir force is that they are vastly different in size, not because the strong Casimir force somehow does not affect electrons.

Note that this theory is based on electrons being the diameter that equates to the mass of the electron due to the displacement of zero-point energy, which is the Compton wavelength of ~2.4 x 10^{-12} meters. For comparison, the radius of the proton has been measured as 0.88 x 10^{-15} meters. For more information about this strong Casimir force between electrons please refer to my paper "Casimir Attraction Between Electrons."[93]

It turns out that in addition to the strong Casimir force, there must be another force. If there was only the strong Casimir force and the electron structure interacts with quantum fluctuations much smaller than its diameter, then the strong Casimir force would become so strong that electrons would never come apart. So, as we have seen with protons, there must be a repulsive force between electrons at very close range. We previously reached the same conclusion when considering electrons in atomic orbits.

Perhaps you are also wondering that since electrons interact due to a type of strong force, why do they not form stable groupings similar to atomic nuclei. The answer to that is simple; there is no electrically neutral particle that is the same size as the electron. Nuclei would not form without neutrons; so similarly, electrons do not form electron nuclei. The strong Casimir attraction between electrons is at best metastable.

The list of fundamental forces is getting even longer. They are the:

1. mechanical force,
2. electromagnetic force,
3. weak force,
4. strong force,
5. dark force,
6. high-energy gravity force,
7. repulsive force between electrons and protons,
8. repulsive force between electrons,
9. repulsive force between protons,
10. repulsive force between protons and neutrons,
11. repulsive force between neutrons,
12. attractive force between electrons.

Here is something else to ponder. When we consider the Casimir effect, it is said that the quantum field pushes against the plates. The deeper question is how do quantum fluctuations push? The answer is; the repulsive forces between electrons, protons and between protons and electrons. Quantum electrons and/or protons push against each other when they get close. The repulsive force between particles is much stronger than the repulsive electrostatic force.

The statement that electrons do not feel the strong force is a lie. They take part in an interaction that is identical to the strong interaction, but only with other electrons.

[93] R. Fleming "Casimir Attraction Between Electrons." Researchgate.net.

Lie #72: Leptons

The 72nd of the greatest lies in physics is leptons. Leptons are defined as half integer spin particles that do not undergo strong interactions. There are six particles defined as leptons: the electron, mu, tau, electron neutrino, mu neutrino, and the tau neutrino.

Right away we can see two big problems with the lepton class of particles. First, five of the six particles are not elementary particles. They should not even be considered particles as they are composed of two or more other particles. The only true particle in the bunch is the electron.

The second problem is that the electron does undergo a strong interaction with other electrons. This strong interaction is identical in principle to the strong interaction between protons. The fact that electrons and protons do not interact with each other, due to their vast size difference, is not sufficient to create a new class of particles.

The lepton class of particles is one of the great lies of physics. **It is not much of a leap to say that lepton conservation is a lie too.** The absence of a strong force between electrons and protons does not make them different types of particles, only differently sized particles.

Lie #73: Protons Are Not Elementary Particles

The 73rd of the greatest lies in physics is protons are not elementary particles. One of the key misunderstandings in physics which sent particle theory in the wrong direction, was the idea that since protons have structure, they cannot be fundamental particles. The thought is, whatever the structure is made of, is what is fundamental. This is fallacious reasoning as it is possible for a particle to have structure and for that structure to only exist as part of the particle, and therefore, the particle and structure are elementary when taken together. **It is a lie to say that a particle is not elementary because it has structure.**

In most places on the Internet you will see a proton made of three quarks in a triangle. When we do scattering experiments that is not what we see. The outside of the proton appears to be spherical or quasi-spherical. The outer shell of the proton appears to be composed of a multitude of smaller particles.

Richard Feynman came up with a model that fit the scattering experiments pretty well. In his model the proton is composed of many smaller particles he called partons.[94] By the word many, he meant thousands or perhaps a 100,000 or more. The proton acts like it is composed of many smaller particles. As the quark theory caught on, they co-opted Feynman's theory and said that the partons must really be quarks, never mind that 3 is a lot different from 100,000.

To make up the difference, quark theorists say there must be lots of gluons, and the quarks must be jumping around a lot, so the proton looks like a cloud of gluons and quarks. They even added in virtual quark pairs to pad the numbers. While physicists have collected a lot more scattering data over the past 50 years, the basic quark-gluon model has not changed much.

But there is a problem; quarks and gluons are not real. The quark model was invented to simplify particle theory, but has only made it more complicated. On top of that, the quark model does not model the most basic resonance, the neutral pion, along with many other mesons. The quark model is a failed theory. Even worse it is numerology.

Gluons are, of course, not real because the gauge boson particle exchange model of force transmission makes no sense. That, and the strong nuclear force can be accounted for as an electromagnetic force interaction, the strong Casimir force, between particles.

Going back to the Feynman model, nobody has ever detected a parton. There is something, however, that numbers far in excess of hundreds of thousands that is known to exist within a proton, and that is quantum fluctuations, and not of the quark variety. Since it is unlikely there are two things numbering in the hundreds of thousands occupying the same small space, Feynman's partons must really be quantum fluctuations.

The biggest clue we have that protons are elementary particles is that they are permanently stable. Protons do not decay on our normal time scales, and based on the scientific evidence to date, the proton half-life is in excess of 10^{34} years.[95,96] **Proton decay is a lie.**

If protons had structure that had an independent existence, protons would decay. If a proton could come apart, it would. Since protons do not decay, their structure must be linked to the proton in such a way that it only occurs within a proton, or a similarly permanently stable particle like the electron.

What is the structure of the proton? Standard model physicists do not know. Since the proton mass equates to a Casimir cavity the same size of the proton, it appears to be a cavity filled with quantum fluctuations which for the spherical charge radius. As Dirac hypothesized about the electron, it looks like a bubble in the zero-point field. It turns out that, like electrons, protons act like bare protons that act like electric and matter-antimatter polarizers surrounded by the polarized quantum field which gives the proton its size, magnitude of electric and matter charge, and its spin, magnetic moment, and mass.[97]

As mentioned in Lie#55, electrons were falsely thought to be small and structureless. Those are obvious lies, as they do not make sense scientifically. One cannot physically describe a particle's mass, charge, matter or antimatter, magnetic field, and/or spin without some type of structure. Once the electron's properties are modeled as quantum field effects, except for a bare central polarizer, it is easy to show that protons are the same only smaller.

The statement that protons are not elementary particles is a lie.

[94] R. P. Feynman, R. P., "The Behavior of Hadron Collisions at Extreme Energies". High Energy Collisions: Third International Conference at Stony Brook, N.Y. Gordon & Breach. pp. 237–249, 1969.
[95] B. Bajc, J. Hisano, T. Kuwahara, Y. Omura, "Threshold corrections to dimension-six proton decay operators in non-minimal SUSY SU(5) GUTs". Nuclear Physics B. 910: 1, 2016, arXiv:1603.03568.
[96] H. Nishino, Super-K Collaboration, "Search for Proton Decay via p+→ e+ π0 and p+→ μ+ π0 in a Large Water Cherenkov Detector". Physical Review Letters. 102 (14), 2012.
[97] R. Fleming, "Electron Properties Explained as Quantum Field Effects." gsjournal.net. 18 Sep. 2018.

Lie #74: Neutrons are not an Electron and a Proton

The 74th of the greatest lies in physics is neutrons are not an electron and a proton. Neutrons are formed when an electron overcomes the potential barrier between it and a proton, and merges with the proton. Then a free neutron will decay to a proton and electron with a mean lifetime of around 15 minutes. Note that neutrons appear much more stable when they are in an atomic nucleus surrounded by protons and other neutrons. They are not actually much more stable since neutrons are continually decaying while new neutrons are being produced.

As with other combination particles or resonances, it is the decay products that give us our best clue to their make-up. Consequently, the best working hypothesis is that a neutron is a combination of a proton and an electron, as they do not decay into quarks or anything else that is detectable. We do need to keep in mind, however, that neutrons are much smaller than a Compton wavelength sized electron, so the electron must be transformed in some way.

Three things got in the way of the electron plus proton model for the neutron: proton structure, lepton conservation, and the quark model. Protons were found to have structure, while electrons were thought to have no structure. After that, physicists were biased toward falsely thinking neutrons were composed of something more elementary too, while at the same time rejecting that an electron was part of that structure.

The second problem was that, since physicists did not understand neutron decay, they invented the neutrino as a way to carry away the excess energy from the decay process. Then they invented a rule based on these unobserved neutrinos, called lepton conservation.

In lepton conservation you must add up the number of leptons on both sides of a decay equation, with antimatter leptons being negative, and make sure that they are equal. As stated before, since leptons are not real, lepton conservation is not real. Nonetheless, the standard decay equation for a free neutron is shown in equation 74-1.

Equation 74-1

$$n^0 \to p^+ + e^- + \bar{\nu}_e$$

In this equation the electron and the anti-electron neutrino cancel so there are zero net leptons on both sides of the equation, and leptons are conserved. Lepton conservation is drilled into physics students so the idea that a neutron was made of a proton plus an electron largely vanished.

There is a strong possibility that neutrinos are their own antiparticle, whether they are composite or not. So how are we supposed to know when to subtract a neutrino or add a neutrino? Think about it. In any case, retaining lepton conservation as a principle is not desirable, particularly since by definition, there are no leptons.

What really happens in neutron decay is that a quantum electron-positron pair interacts with the neutron. The virtual positron annihilates with the electron-like part of the neutron leaving behind a proton and a once virtual but now free electron. This is how an electron tunnels through the potential barrier, by not really tunneling at all. This equation is shown in Equation 74-2.

Equation 74-2

$$n^0 + (e^- e^+) \to p^+ + e^-$$

In reality it looks like the electron does not go away when it is inside a neutron, it is just transformed in some manner. Physicists have come up with other made up rules that get in the way of doing real physical particle physics, but we will not go into them. It is enough to know that particles do not know the rules, they act based on their physical restrictions

As for the third problem, nothing additional really needs to be said about quarks. It is a false numerological model that should be discarded.

That leaves us with the concept of a neutron is a compound particle composed of a proton and an electron. We can imagine how that may be. Since the size of a neutron is similar to a proton, the neutron follows the same mass rule as a proton. Its mass is determined by the zero-point energy it excludes when treated as a Casimir cavity. That means that the electron must shrink in size, with its outer quantum shell disappearing when a bare electron gets inside the proton's outer quantum shell. Figuring out exactly how that works needs to go on someone's to-do list.

The structures of all three particles, the electron, proton, and neutron, are filled with quantum fluctuations. There must be something about their structures that allows us to see the particles as something of a collection of quantum fluctuations when we do scattering experiments. The particle structure causes them to interact in the way Feynman described in his parton model. Something about the structure makes the quantum fluctuations detectable, sort of like how a balloon makes the gas inside detectable.

The hypothesis that neutrons are not a proton plus and electron is a lie. While consideration of neutron structure is speculative at this point, it is clear that going forward, it is necessary to treat neutrons as combinations of protons and electrons and see where that leads us.

Lie #75: Particles have Relativistic Mass

The 75th of the greatest lies in physics is particles have relativistic mass. Relativistic mass is the concept that things gain mass-energy as they increase velocity. This is a relativistic effect like clock slowing, and the same term is used in the denominator of the rest mass or energy equation (Equation 75-1) as we saw earlier for distance and clock rates.

Equation 75-1

$$E = \frac{mc^2}{\sqrt{1 - v^2/c^2}}$$

A troubling question about this is what does this mean to a particle's structure? And, does the particle's mass really increase, or is it something else? The aether deniers are stuck here, as without zero-point energy around a particle, they were stuck with the idea that particles actually get heavier.

Even worse, since they believe in length contraction (Lie #14) they were faced with the idea that any particle with physical dimensions would be shorter in the direction of the velocity, so if it started out spherical, it would end up shaped like a pancake as it neared the speed of light.

Then we have the question, does the particle get bigger or smaller? Ultimately, they decided to treat particles as points (Lie #54) in order to avoid these structural questions. If you really want to make your head swim, try to figure out particle structure using special relativity, when there is no standard reference frame, and no way to tell which observer should be seeing a spherical particle or a pancaked particle.

It turns out there is another solution, and it requires aether. If we recall the description of inertia from Lie #22, inertia is due to the rotation of quantum fluctuation dipoles. This rotation leads to the formation of a field that is similar to the magnetic field around a moving charge, but in this case, it is due to interactions with the mechanical dipole, matter and antimatter. The relationship between this magnetic-like mechanical field and the moving body of matter follows the rules of self-induction. Movement of the body causes the field to form and the field causes the body to move.

Half of the kinetic energy of the moving body is in the field, although it is traditionally ignored in physics math. The kinetic energy formula is $\mathbf{E = \frac{1}{2}mv^2}$, while in truth it should be just $\mathbf{E = mv^2}$. That said, as long as we consistently ignore half the energy, the math works out just fine. Note that the same logic was used originally to justify the equation $\mathbf{E = mc^2}$ for rest mass. Based on the inertia model, all the inertial and kinetic energy are actually in the quantum field.

When it comes to the mass of the particle, we have something similar. The particle is only truly at rest when it is at rest in the aether rest frame. In all other frames of reference, it has a velocity relative to the aether, and that velocity leads to an increase in the energy of the magnetic-like mechanical field. The magnetic-like mechanical field energy is the relativistic mass-energy.

'Relativistic mass' is in the field, not the particle, and it is energy not mass. We can also recall that length contraction is a lie, so we do not have to be concerned with a physical pancake effect, it simply does not happen. Particle structure is always the same, regardless of the particle's velocity. And, since particle mass is also zero-point energy, all mass, relativistic or otherwise, is zero-point energy.

The idea that particles have relativistic mass is a lie. It is the aether, the zero-point field, that carries the extra energy that we incorrectly refer to as 'relativistic mass.' **To that end, the term relativistic mass is a lie too.** It should just be called relativistic energy.

Lie #76: Mass is due to the Higgs Field

The 76th of the greatest lies in physics is mass is due to the Higgs field. Peter Higgs, François Baron Englert and Robert Brout came up with the idea that mass comes from a new type of field, now called the Higgs field, with this new field being due to the Higgs boson.[98,99] This work was important as no standard model physicist aside from Dirac had the answer to the question; where does mass come from? As discussed in lie #19, mass is not intrinsic. This was particularly important to weak interaction theory, as it was unknown how the W and Z particles could end up with so much mass. After the discovery of the bump in data identified as the Higgs Boson was announced, Higgs and Englert won the Nobel Prize.

The Higgs hypothesis is a familiar case of physicists' standard modus operandi of inventing a new dimension, field, and/or particle when they do not know how something physically works. Physicists had the zero-point field, the electro-magnetic fields, the photon field, the neutrino field, the gluon field, the graviton field or gravitational waves, not to mention the cosmic background radiation as another kind of field. Why not invent a new field to account for mass?

In prior chapters it has been shown that all the other fields in physics reduce to the zero-point field, so it is not hard to imagine what happens to the Higgs field, particularly since there is a better explanation for the origin of mass. The mass of electrons, protons, and neutrons is not due to a Higgs field. Those particle masses are due to those particles excluding zero-point energy equal to their masses. Look at Lie #19 and the referenced paper for more details.[38]

The unstable particles, more properly called resonances, are somewhat different. Basic Sternglass models for the pion, mu, and tau are shown in Lies #59, #62 and #63. As with the electron and proton, their mass is equal to the amount of zero-point energy they exclude, however, that exclusion is not all in the form of a spherical cavity but rather, relativistic mass.

The mass of resonances comes in multiple parts. While quantum fluctuations do not have mass, if a quantum fluctuation exists longer than allowed by the Heisenberg uncertainty principle, the constituent particles must have their normal rest mass. Consequently, the electrons and positrons in the pion, mu, and tau all have their normal rest mass of 511 keV/c^2. This rest mass is due to their Casimir cavity.

Because the Sternglass model is a relativistic model with the constituent particles moving faster than the speed of light, they also have relativistic mass. Relativistic mass is, however, not due to a change in the dimensions of the particle and its Casimir cavity. Relativistic mass is inertial energy trapped in the rotating quantum fluctuations of the inertial zero-point field.

Smaller mass-energy contributions due to angular momentum and magnetic fields are similarly due to the energy of the quantum fluctuations that make up those fields.

The masses of all resonances are likewise due to a combination of the rest mass of the temporarily excited particles, along with their relativistic mass, other inertial masses, and mass-energy due to the resonance's fields. Each of these sources of mass-energy displaces zero-point energy. All forms of mass-energy of resonances and particles are due to the displacement of zero-point energy.

Saying mass is due to the Higgs field is a lie. Mass is not due to the Higgs field; it is due to the zero-point field.

[98] F. Englert, R. Brout, "Broken Symmetry and the Mass of Gauge Vector Mesons." Physical Review Letters. 13 (9): 321–323, 1964.
[99] P. W. Higgs, "Broken Symmetries and the Masses of Gauge Bosons." Physical Review Letters. 13 (16): 508–509, 1964.

Lie #77: Higgs Boson

The 77th of the greatest lies in physics is the Higgs Boson. There has been a lot of science news in recent years about the large hadron collider experiment that was used to search for the hypothetical Higgs boson. Physicists announced that it was found in July 2012. The discovery of the data bump they called the Higgs boson is thought by mainstream physicists to confirm the existence of the Higgs field and solve the origin of mass question. They were of course wrong as usual.

What physicists discovered was a barely statistically significant bump in their data at approximately 126 GeV energy. This bump had no electric charge, 0 spin, and other properties that matched the hypothetical Higgs boson. It basically has all the properties you would expect from a resonant quantum fluctuation; in other words, nothing distinguishing. It also has the mass-energy of a single relativistic proton being essentially equal to the proton mass divided by the fine structure constant. In an odd break with physics experimental protocol of the past, the Higgs boson was accepted without another laboratory confirming it independently.

There have been papers and even a book written about the lack of scientific integrity in the process of the Higgs discovery. Dr. Alexander Unzicker's book *The Higgs Fake – How Particle Physicists Fooled the Nobel Committee* is highly recommended as a good summary of the problems related to the Higgs experiment. He is not even certain that the statistical treatment of the data is valid, as no one has performed an independent analysis of their data. Dr. Unzicker is also justifiably critical of the quark theory and the scientific establishment in general.

That said, the primary reason to dismiss the Higgs boson discovery is that the Higgs boson, along with the Higgs field, are unnecessary. Masses of particles are due to the zero-point field. Mass is equivalent to the zero-point energy excluded by a particle or resonance. The Higgs hypothesis is simply not needed.

You may ask then, if we assume that this statistically barely significant bump in highly manipulated data is real, and is one day confirmed, what is it? As with all resonances, the best way to determine what it is made of is to look at its decay products. In the Cern experiments, the Higgs boson was thought to decay into one of four possible options:

    a. a pair of photons,
    b. a tauon and anti-tauon particle pair,
    c. a W^+ and W^- particle pair,
    d. a Z boson pair.

Note that the literature mentions quark pairs, but the experiments do not actually detect quarks, so we will only consider the real physical decay products and the W and Z resonances. Note that, while decay modes (c) and (d) violate the principle of conservation of energy they point to the Higgs being related to the W and Z and having the same cause of their mass. The mass of the Z is about 3/4ths the mass of the Higgs which is similar to the ratio between a muon and pion. The cause of the W, Z, and Higgs mass, relativistic protons, was discussed in chapter 68 and the paper in citation 91. The mass of the proton times the inverse of the fine structure constant (~137) equals ~128 GeV/c^2, which is very close to the known Higgs mass.

The Higgs boson is a lie, and **the Higgs field is a lie as well**. Neither hypothesis was ever actually needed.

Lie #78: Bosons are Elementary Particles

The 78th of the greatest lies in physics is bosons are elementary particles. To anyone keeping a running tally in their head, they already realize that every elementary boson in the standard model is a lie. There are still, however, compound particles and atoms that behave like bosons, and follow Bose-Einstein statistics, but they are not elementary particles. The standard model's list of elementary bosons includes:

   a. photons – Lie #4
   b. gravitons – Lie #49
   c. W & Z bosons – Lie #68
   d. gluons – Lie#70
   e. Higgs Bosons – Lie #77

Photons are a series of quantum dipoles in combination with a field of quantum fluctuations that make up its wave. Even though they are composite particles, photons are still important energy carriers.

Gravitons are fictional. High-energy gravity is a Fatio-Casimir force consistent with van der Waals forces in electromagnetic and mechanical force theory.

W and Z bosons are composite resonances containing a relativistic proton. Due to their mass, they cannot act as gauge bosons without violating the principle of conservation of energy. The W and Z particles apparently serve no real purpose.

The strong nuclear force is easily accounted for as a strong Casimir force that is consistent with electromagnetic theory. It has nothing to do with the fictional gluon.

The Higgs boson, if it actually exists, also appears to be a composite particle containing a relativistic proton. This resonance serves no purpose, as it is unnecessary to explain the origin of mass.

From Lie #18 we can also recall that the entire gauge boson, smart exchange particle hypothesis is fiction. It simply cannot account for force transmission and acceleration due to forces.

The statement that bosons are elementary particles is a lie. All of them are either composite particles or purely fictional.

Lie #79: The Table of Elementary Particles

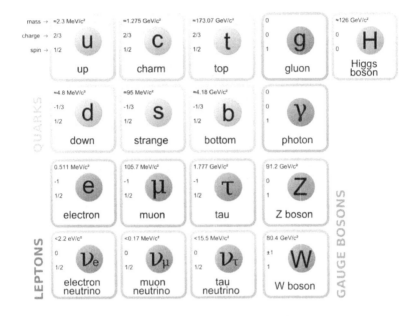

Figure 79-1 Table of Elementary Particles in the standard model of physics[100]

The 79th of the greatest lies in physics is the table of elementary particles. Just look at the table above. Where do we start? Honestly, how wrong can a model, and many thousands of physicists be? There is only one permanently stable particle, one particle that is truly elementary in the whole table. The rest are either resonances, or purely fictional.

The categories by themselves, quarks (Lie #61), leptons (Lie #72), gauge bosons (Lie #78) are each lies on their own. The resonances called quarks and gauge bosons do not perform their stated function, or are fictional. Individually, the gauge bosons are lies, as addressed in the preceding chapter.

The neutrinos (Lie #64 and Lie #65) as a group are a lie, as each is functionally either a quantum particle pair, or a series of quantum particle pairs left over from an interaction. They do not have mass just like photons do not have mass (Lie #66).

The mu and tau are not elementary particles either (Lie #62 and Lie #63). There has been a model in existence for the mu for over 50 years that is widely ignored by standard model physicists, and the tau is readily shown to be two-kaon onium resonance. Those models need tweaking, but the mu and tau are clearly not fundamental.

After all those lies, physicists then miss a particle due to the false assumption that a particle with structure cannot be elementary. Every particle, that actually is a particle, and has the usual particle properties, cannot be a point particle (Lie #54). Consequently, they must have physical dimensions, and they must have structure. The fact that the proton's mass is equivalent to the zero-point energy excluded by a spherical structure of its known size proves this. The proton, the only other permanently stable particle beside the electron, is elementary (chapter #73).

A fundamental theory of physics only requires two elementary particles, the electron and proton, plus their antimatter analogs, the positron and antiproton.

The standard model table of elementary particles is a lie. We could make the actual table, but an elementary particle table with two particles seems silly.

[100] Public Domain image by: MissMJ - Own work by uploader, PBS NOVA [1], Fermilab, Office of Science, United States Department of Energy, Particle Data Group)

Lie #80: Copenhagen Interpretation

The 80th of the greatest lies in physics is the Copenhagen interpretation. A collection of the greatest lies in physics would not be complete without addressing some of the hypothetical ideas that have come out of the statistical uncertainties of the quantum mechanical model of electrodynamics. As with so many of the lies, this one comes down to physicists ignoring aether.

The Copenhagen interpretation is a school of thought with regard to how to interpret results in quantum mechanics. The pioneers of this informal school were Neils Bohr and Werner Heisenberg who both did research at the University of Copenhagen.[101,102]

The interpretation is in line with Heisenberg's statement of his uncertainty principle and the related observer effect. In his principle, certain pairs of properties, such as position and momentum, cannot be known at the same time. The observer effect states that measurements cannot be made without affecting the system. Hence if we measure one of the two complimentary properties in a quantum system the other is unknown.

The Copenhagen interpretation takes the observer effect a step further and says that in a quantum system with probabilities for more than one result, the act of observing, or in particular measuring, makes one of the results happen. Note that the observer in this case can be any outside classical system, although this classical system can act in a manner usually thought of as something that happens in quantum theory.

Another way it is stated is that a quantum system is described by a wave equation that yields a distribution of probabilities of different results happening. When the system is measured, the interference causes the wave equation to collapse yielding a single result. Note that scientists have argued indefinitely on what the Copenhagen Interpretation really means, including Bohr and Heisenberg, so feel free to study it on your own and think of your interpretation going forward.

Einstein and others have challenged the Copenhagen interpretation over the years, and to paraphrase Einstein, the universe does not play dice. Einstein, as an aether denier, however, did not allow himself to use the tools needed to challenge it properly.

One of the competing interpretations is the De Broglie-Bohm theory.[103] In that theory a configuration involving all particles actually exists, even though they are unseen, and those particles and fields determine what the result will be. That sounds a lot like they are describing the zero-point field. In the De Broglie-Bohm model they still consider that this unseen field of particles can be described by a still larger more all-encompassing wave equation.

We do know that the zero-point field exists. The zero-point field can in principle be described by a greater all-encompassing wave equation. It would be necessary to do so in order to have a comprehensive model, albeit a very challenging task to accomplish.

It is interactions with individual quantum fluctuations that ultimately determine what happens, in the manner described in chapter 57. Quantum particle pairs interact with particles, and photons interact with particles by way of annihilation-production interactions. While quantum fluctuations are also probabilistic in terms of their distribution, results happen due to the existence of individual quantum fluctuations interacting at a given point and time.

Outside interference, such as from a measuring apparatus, is unnecessary. It is not necessary since there are quantum fluctuations interacting with particles all the time. While it is certainly true that a measurement can interfere with a quantum system and cause a certain result, a result can be achieved without any outside interference.

The Copenhagen interpretation is a lie. Quantum fluctuations exist and interactions with quantum fluctuations determine the outcomes.

[101] N. Bohr, "The quantum postulate and the recent development of atomic theory." Nature, 121: 580–590, 1928.
[102] W. Heisenberg, "Language and reality in modern physics." Chapter 10, pp. 145–160, in *Physics and Philosophy: the Revolution in Modern Science*, George Allen & Unwin, London, ISBN 0-04-530016 X, p. 153, 1959.
[103] D. Bohm, "A Suggested Interpretation of the Quantum Theory in Terms of 'Hidden Variables' I." Physical Review. 85 (2): 166–179, 1952.

Lie #81: Spooky Action at a Distance

The 81st of the greatest lies in physics is spooky action at a distance. When arguing against the Copenhagen interpretation, Einstein along with Boris Podolsky and Nathan Rosen came up with what is known as the EPR paradox.[104] They invented the concept of quantum entanglement. Two particles can be entangled in a way that they have a property that complements each other. A common property like this is spin, where the two particles—Fermions to be specific—in an otherwise identical state have opposite spin orientation.

One version of the argument goes, that if two particles are entangled and become separated and you measure a property such as spin for one particle, then you will instantly know the property of the other. Further, under the Copenhagen interpretation, if the first particle's property, such as spin state, is not known until the measurement is made, then the property in the second particle must be set instantly, even if they are at a distance from each other. This effect violates the hypothetical speed of light limit for information. Einstein dubbed this type of quantum communication spooky action at a distance.

It has been found in numerous experiments that quantum entanglement does lead to spooky action at a distance. There are many papers, popular science magazines, and blog articles that report these phenomena. Nonetheless, supporters of both the Copenhagen interpretation and the speed of light limit for information travel pretend that it does not matter.

Now you may be thinking, if it has been confirmed, how is it a lie? Well, the lie is threefold. The obvious one being that it is not in any way spooky. It is also not action at a distance, and not a violation of the principles of physics. For one, the property may have been set by an interaction with the zero-point field long before the measurement occurred.

Recalling Lie #38, electric and magnetic fields propagate at the speeds much faster than the speed of light, so two particles that are quantum entangled are in direct communication in a way that allows for transmission of some basic information at a speed much greater than that of light. This is not spooky.

There is no action at a distance going on as the information is being transmitted through the zero-point field. There is also no violation of a velocity limit, since basic information like electric and magnetic fields and, now we know, information about spin states can propagate at speeds much greater than the speed of light. Of course, the information may not need to travel faster than the speed of light at all, if a zero-point field interaction determined the particle states before they moved apart.

Spooky action at a distance is a lie. This is normal physics. All one needed to do was to make sure to include aether in the theory and understand a little about force transmission.

[104] A. Einstein, B. Podolsky, N. Rosen, "Can Quantum-Mechanical Description of Physical Reality be Considered Complete?" Physical Review. 47 (10): 777–780, 1935.

Lie #82: The Many-Worlds Interpretation

> *And in Dublin in 1952 Schrödinger gave a lecture in which at one point he jocularly warned his audience that what he was about to say might 'seem lunatic'. It was that, when his equation seems to be describing several different histories, they are not alternatives but all really happen 'simultaneously'.* [105]
>
> David Deutsch, 2011

The 82nd of the greatest lies in physics is the many-worlds interpretation. The many-worlds interpretation has been getting a lot of attention in popular science articles and televised programs as well as science fiction, and is most often called the multiverse theory. It is an alternative idea to the Copenhagen interpretation based on the idea that instead of only one probable result happening, they all happen, and each different result leads to the formation of a new world or universe.

As the above quote shows, Erwin Schrödinger was the first to propose the idea, while the first paper published was by Hugh Everett in 1957, who came up with the idea independently.[106,107] However it was not until Bryce DeWitt started publicizing the many worlds interpretation that the idea started to take off.[108,109,110]

There are almost as many multiverse models as there are scientists and science fiction writers who support the idea. Most of the ideas are so wild that it is impossible to tell the difference between science fiction writers and scientists. This is one of the worst cases of the real science of physics being damaged by science fiction. The multiverse idea brings in so much book and television revenue, because it is honestly quite interesting to think about and very popular with the general public, but the damage it is doing to the integrity of real physics is incalculable.

The first problem is that each new universe that would be created is a massive violation of the principle of conservation of energy. The second is that there is no space in which to put it. It is true that physicists use the concept of extra dimensions to find a place for it, and that is a lie that will be dealt with in a later chapter.

The number of new universes created within the scope of a many worlds interpretation is too great to imagine. Think for the moment of an electron in orbit around a carbon nucleus. It is not really in orbit, but it fills a cloud like region of space, where the cloud's dimensions are described statistically. The electron does not move in a continuous orbit. Instead the electrons make discreet little jumps from place to place, quantum jumps as they are called.

Mainstream physicists cannot describe how electrons make these quantum jumps, so they assume all probabilities are possible and must be accounted for in some way. Their problem once again goes back to aether denial. Once we recognize that the space around the electron is filled with aether, and in particular quantum electron-positron pairs, it becomes obvious how electrons jump as described in Chapter 57.

When an electron and a positron come together, they annihilate one another. But a quantum electron-positron pair cannot only annihilate with the other half of its pair; it can cross-annihilate with another quantum electron-positron pair or even free electrons. Electrons do not really jump, but rather a new electron forms at some distance away.

Now depending on how one measures the positions in a probabilistic cloud occupied by an electron, there is potentially an infinite number of possible positions. But, say for example we divide the cloud region into a cube with 100 units to a side giving a million possibilities. Then say the electron jumps a million times a second.

If a new universe is created for each of the million possibilities each time this one electron jumps, a trillion new universes would have to be created each second for the many-worlds interpretation to be valid for just a single electron.

Now multiply that by the 10^{80} electrons estimated to be in a big bang type universe, taken for the sake of argument, and then multiply that by 13.8 billion years, once again for the sake of argument, then double that to include protons, and you get about 10^{110} universes that need to be created. That is impossible. The many-worlds interpretation is obviously a ridiculous assertion.

To add to that, the entire premise is incorrect, as the other probabilities no longer exist once the electron makes the jump. The jump is a real physical process mediated by a real physical quantum fluctuation. While the quantum fluctuation's existence can be described probabilistically, once it exists and the electron jumps, it is over, until the next time. The other results cannot happen. As quantum physicists say, the wave function collapses at that point to a single result. The other probabilities are no longer possible.

Every other possibility described by quantum probabilities is similarly deterministic. A quantum fluctuation, photon, electron, or other particle is in the right place at the right time for a specific result to occur. The other results simply do not occur. As mentioned in Chapter 80, this is not consistent with the Copenhagen interpretation or the many-worlds interpretation.

Lastly, we can consider the argument that an alternate world or multiverse can never be detected, as they could only exist outside our universe. There is little point to speculating about nonsense. This is like philosophy or religion, not science.

As with the non-space hypothesis, the many-worlds and multiverse hypotheses are non-scientific since:

A. There is no physical evidence for them.
B. They are unnecessary to describe the universe.
C. They add unnecessary new complications to the theories of our universe.
D. They can never be detected.

The many-worlds interpretation and the related **multiverse models** are all lies. There is one universe.

[105] D. Deutsch, *The Beginning of Infinity*, page 310, 2011.
[106] H. Everett, "Theory of the Universal Wavefunction." Thesis, Princeton University, (1956, 1973), pp 1–140
[107] H. Everett, "Relative State Formulation of Quantum Mechanics." Reviews of Modern Physics. 29: 454–462, 1957.
[108] B. S. DeWitt, "Quantum Mechanics and Reality: Could the solution to the dilemma of indeterminism be a universe in which all possible outcomes of an experiment actually occur?" Physics Today, 23(9) pp 30–40, September 1970.
[109] B. S. DeWitt, "The Many-Universes Interpretation of Quantum Mechanics." Proceedings of the International School of Physics "Enrico Fermi" Course IL: Foundations of Quantum Mechanics, Academic Press (1972)
[110] B. S. DeWitt, R. N. Graham, eds, *The Many-Worlds Interpretation of Quantum Mechanics*, Princeton Series in Physics, Princeton University Press (1973).

Lie #83: The Wave Model of Quantum Mechanics

> *Planck's formula tells us that the kind of "field" that actually occurs in nature has some sort of discreteness about it, which makes it behave like a system of particles, where the higher the mode of frequency of oscillation that the field might indulge in, the more strongly the energy in the field would manifest itself in this particle-like behaviour.*[111]
>
> Roger Penrose, 2016

The 83rd of the greatest lies in physics is the wave model of quantum mechanics. Wave models are the basis of quantum mechanics, and quantum mechanics uses wave equations to describe practically everything at the quantum scale. The first wave equation is the Schrödinger equation, which describes the hydrogen atom. A second famous wave equation is the Dirac equation that describes electrons.

As discussed in the previous few chapters, quantum mechanical probabilities are based on possible results described by the applicable wave equation. The result is said to occur when the wave equation collapses. So, anyone who studies quantum theory will see 'wave' this, and 'wave' that, throughout their course of study. From a quantum mechanical perspective, we are led to believe that the wave properties are what is fundamentally real.

The wave model evolved due to the rejection of aether, which made it a requirement that a particle be both a particle and a wave. As discussed way back in Lie #3, the wave-particle model is a lie. The wave is a property of aether, not a property of the particle. Furthermore, aether consists of particle dipoles, so even the waves are really very large numbers of particles.

Waves cannot have point-like electric charges. The existence of electric charges or polarizers and electric charge dipoles requires that there be particulate matter. And hence, the quantum field must contain pairs of particles that form dipoles rather than waves which cannot form dipoles by themselves.

Now those quantum fluctuation particle pairs act like waves when evaluated in large numbers, and the macroscopic properties of the zero-point field can be described by a wave function. Quantum mechanical wave equations serve their mathematical function and allow us to make calculations, but at the microscopic scale, aether is still a bunch of particles.

This is one of those occasions where mathematical physicists want to believe their equations are a fundamental description of physical reality, when they are not. The wave equations are nothing more than a mathematical tool. They are not consistent with the physical description of what is going on within the aether.

Even though particles can be described by a wave equation when we conduct scattering experiments, they behave like a volume of space filled with a bunch of particles, and those particles are quantum fluctuations, which are particles. It's particles all the way down.

The wave model of quantum mechanics is a lie. It would be better if physicists treated wave equations as what they are, useful mathematical tools. They are good tools, as writing a particle-based equation is difficult when you are dealing with essentially an infinite number of particles. But they are still tools, not an elementary physical description of nature.

[111] Roger Penrose, *Fashion Faith and Fantasy in the New Physics of the Universe*, pg 192, Princeton University Press, 2016.

Lie #84: The Kaluza-Klein Theory

The 84th of the greatest lies in physics is the Kaluza-Klein theory. Unless you have studied a lot of physics, you may not be familiar with the Kaluza-Klein theory, but it was an early attempt at unifying two forces and is a precursor of sorts to string theory.[112,113,114] It was surpassed by string theory, so the physicists in the crowd are probably thinking this is old news.

The importance of the Kaluza-Klein theory is that it introduced three new methodologies for attempting to unify two or more forces. The first of these methodologies is adding a new dimension to a theory to account for a force. In this case it was electromagnetic theory being added to general relativity, so it turned 4-dimensional general relativity model into a 5-dimensional theory that included electromagnetism. The 4 dimensions of general relativity include the 3 physical dimensions plus time which is treated as an extra dimension.

The second new methodology was, instead of merely treating the new dimension as a mathematical construct, they decided that it is physically real. In practice, having an N-dimensional problem-solving space in order to solve an N-variable problem is OK. Thinking that all of those N dimensions are physically real is where they get into trouble.

The third methodology is that when faced with having a new physical dimension that is not real, they decided that it must be curled up, so it is somehow hidden from view. A common example of how this is supposed to work is the garden hose model. From a distance a garden hose looks like a line, but up close it looks like a cylindrical tube. The extra physical dimension in the Kaluza-Klein theory is sort of curled up like that, so we only see it is a line.

Of course, it is nice that we can combine two force theories into a single set of mathematical equations. Even treating one of those force theories as a new mathematical dimension is a reasonable mathematical approach to problem solving. **It is a lie to say this new physical dimension, which does not actually exist as a physical dimension, is somehow curled up so we do not see it**. This is why we must be careful when we have mathematicians pretending they are physicists.

The Kaluza-Klein theory predates the weak force and strong force, so it became clear that if someone wanted to use this approach to invent a unified force theory, they would have to invent even more invisible dimensions. We will look at that in the next chapter.

In the end we do not need to invent a new dimension to combine gravity and electricity and magnetism. The Fatio-Casimir effect from electromagnetic and mechanical force theory does a good job accounting for gravity. Gravity is a purely electromagnetic phenomenon. The Kaluza-Klein theory is an unnecessary complication rather than a simplification of our force theories.

The Kaluza-Klein theory is a lie. It is an unnecessary and unnecessarily complicated approach to combining gravity and electromagnetism into one theory.

[112] T. Kaluza, "Zum Unitätsproblem in der Physik." Sitzungsber. Preuss, Akad, Wiss, Berlin. (Math. Phys.): 966–972,1921.
[113] O. Klein, "Quantentheorie und fünfdimensionale Relativitätstheorie." Zeitschrift für Physik A. 37 (12): 895–906, 1926.
[114] O. Klein, "The Atomicity of Electricity as a Quantum Theory Law." Nature. 118: 516, 1926.

Lie #85: String Theory

> *Physicists tend not to be over-worried by detailed matters of mathematical and ontological inconsistency if the theory, when applied with appropriate judgement and careful calculation, continues to provide answers that are in excellent agreement with the results of observation – often with extraordinary precision – through delicate and precise experiment. The situation with string theory is completely different from this. Here there appear to be no results whatever that provide it with experimental support.*[115]
>
> Roger Penrose, 2016

The 85th of the greatest lies in physics is string theory. In string theory, instead of particles being thought of as point particles, they are thought of as strings. The properties of particles, such as charge, mass, spin, are then thought of as properties of the string. It is in essence a contorted way to attribute properties to a point particle, whereas a point particle should not have any properties, or even be a particle.

The leading problem with string theory is there is no such thing as a string. **The concept of a string as a particle model is a lie.** In addition to that, the most popular string theories have 10 or 11 dimensions, and gasp; the mathematicians who came up with string theory have the gall to think those extra dimensions are real.

Then they borrowed a page from the Kaluza-Klein theory playbook, and decide that each of these extra dimensions must be curled up in some fashion so we simply do not see them. Note that we skipped a lot of stuff to get to the worst parts, so it is left to the reader to study string theory if they want to study the numerous, slightly more subtle mistakes.

String theory branched out into a variety of different theories. **There was supersymmetric string theory also called superstring theory, along with several other types. These were eventually combined into M-theory. Then they added D-Branes to do more Kaluza-Klein like compaction of the hypothetical 11th dimension. These are lies, all lies.**

As the scientific evidence from scattering experiments shows, protons and electrons have physical dimensions. As the mass derivation shows, these dimensions are consistent with mass and particles' ability to gravitate. Their dimensions are also consistent with their magnetic moments.

As discussed previously it is a simple matter to show that the magnitude of charge and matter, spin, angular momentum, magnetic moment, and mass of electrons and protons can be explained as quantum field effects. We only need to assume the polarized quantum field forms a spherical structure at the proton charge radius and an equivalent electron structure at a diameter equal to its Compton wavelength.

There is also the issue, as will be discussed in Lie #100, that the forces are readily unified within electromagnetic and mechanical force theory. There really is no reason to add extra dimensions to account for forces which already should be considered part of the combined electromagnetic and mechanical force theories.

The saddest thing is the waste in human and fiscal capital in the pursuit of string theory. Billions of dollars and more than a thousand physicists' careers have been spent in pursuit of nonsense. Many rationally minded and brilliant young physicists have been detoured from real physics research and become stuck working on this irrational theory. Others, such as myself, chose career paths outside physics academia rather than see their ideas drown in a sea filled with irrational thinking physicists.

String theory is a lie. Today, many of the brightest minds in physics are willing to admit that string theory is not real, never was real, and never will be real. It is time for the physics community to spend its time and funding on better theories.

[115] Roger Penrose, *Fashion Faith and Fantasy in the New Physics of the Universe*, pp 2-3 Princeton University Press, 2016.

Lie #86: Extra Physical Dimensions

The 86th of the greatest lies in physics is extra physical dimensions. We live in a world with three physical dimensions. If you look forward and describe these dimensions in the usual way, one dimension is up and down, another side-to-side, and a third is the distance away from you. Those are convenient linear coordinates that we use to locate the every day things we see. There are of course other three-dimensional coordinate systems we can use, but three physical dimensions is reality. Extra dimensions are not physically real.

Empty space, if there were such a thing, would be dimensionless, as it would have nothing in it to measure distance. Real space contains zero-point energy, and quantum fluctuations have wavelengths and frequencies. It is the wavelengths of the quantum fluctuations that give space dimensions. It is the frequencies that give space its property of time.

Wavelength and frequency are not the same thing, so physical lengths and time are not the same thing either. The mainstream model includes a four-dimensional space-time that combines the three physical dimensions and time by treating them as four-dimensions, but time is still something distinct from the three physical spatial dimensions, just as frequency is distinct from wavelength. **Someone stating that time is a physical spatial dimension is lying**.

It is also important to remember that aether has a rest frame where quantum fluctuation properties are average in all directions. In an aether rest frame without stable matter, physical spatial dimensions and the physical clock rates would be uniform throughout the universe. Aether by itself is dimensionally flat and Euclidian with no extra dimensions.

We even have physical evidence for the uniformity of the aether rest frame. The CMB is highly uniform in every direction which confirms the uniformity of the aether, including its wavelengths and frequencies, and thus its physical dimensions and clock rate.

Kaluza's five dimensional theory, with four physical dimensions, began a trend among physicists to add new dimensions to their mathematical models to account for additional variables, such as forces or particle properties.[116] Inventing a many dimensional problem solving space is something that mathematicians are known to do when they have a many variable problem. Once again, this is OK if you are only using it to solve mathematical problems.

But what is not OK, is inventing new mathematical dimensions when they are not even necessary to solve the problem. Since we can combine forces under a combined electromagnetic and mechanical force, extra dimensions are unnecessary to unify forces and furthermore, complicate the theory unnecessarily.

These newly invented dimensions are not physically real, never have been physically real, and never will be physically real. Even if they could exist, they would be outside our physical universe and undetectable. An extra dimension hypothesis is not even wrong.

Note that we can tell we are facing a religious hypothesis rather than a scientific one when:

    A. There is no physical evidence for it.
    B. It is unnecessary to describe the universe.
    C. It adds unnecessary new complications to the theories of our universe.
    D. It can never be detected.

Extra physical dimensions are lies. They are not real, not necessary, and not even scientific.

---

[116] T. Kaluza, "Zum Unitätsproblem in der Physik." Sitzungsber. Preuss. Akad. Wiss. Berlin. (Math. Phys.): 966–972, 1921.

Lie #87: The Speed of Light is Fundamental

> *There will be a physics in the future that works when hc/e^2 has the value 137 and that will not work when it has any other value. (...) The physics of the future, of course, cannot have the three quantities h, e and c all as fundamental quantities. Only two of them can be fundamental, and the third must be derived from those two. It is almost certain that c will be one of the two fundamental ones. The velocity of light, c, is so important in the four-dimensional picture, and it plays such a fundamental role in the special theory of relativity, correlating our units of space and time, that it has to be fundamental.*[117]
>
> Paul Dirac, 1963

The 87th of the greatest lies in physics is the speed of light is fundamental. The quote from Dirac above speaks to an important goal in physics, the reduction in the number of fundamental constants. The value 137, of course refers to the inverse of the fine structure constant. He felt that, along with the fine structure constant, there are four important constants in the equation, but only two are fundamental constants.

As an example, the largest group of fundamental constants in the standard model is the masses of all the particles, currently numbering 122 and counting. Since there is no standard model theory to physically explain those masses, mainstream physicists are stuck with intrinsic or fundamental masses. As discussed in Lie #19, intrinsic mass is a lie. In any acceptable theory, all of those presently fundamental constants must be derived from more fundamental properties.

The Dirac and Sternglass models for mass provide solutions to this problem, vastly reducing the number of fundamental constants. The current non-theory of masses is untenable. The Sternglass model also tells us that the inverse of the fine structure constant is equivalent to the energy of a relativistic electron-positron pair in MeV. This is an important clue that should help us determine the physical relationship between these four constants.

As Dirac pointed out, the speed of light is considered almost sacrosanct as a fundamental constant. To true believers of the special and general relativity hypotheses the constancy and intrinsic nature of the speed of light is untouchable. If the speed of light is not a fundamental constant, both of those theories are invalid.

As stated before, their wrong mindedness goes back to the rejection of aether. If there is no aether, then light has no medium of transmission. If light has no medium of transmission, then the speed of light is an intrinsic property of light. But light does have a medium of transmission, so it is a lie to say that the speed of light is intrinsic to light. The speed of light is a property of the medium.

To see where the intrinsic speed of light theory runs into problems, we only need to look at the equation for the speed of light in terms of the electric constant (vacuum permittivity) and the magnetic constant (vacuum permeability).

Equation 87-1

$$c = \frac{1}{\sqrt{\varepsilon_0 \mu_0}}$$

One can see right away that the electric and magnetic constants should be considered more fundamental than the speed of light. The electric constant relates to the aether's ability to permit the formation of electric field lines. The ability of aether to form electric field lines and the constraints on that property are due to physical properties of quantum fluctuations. The electric constant is not a fundamental constant. **To state that**

the electric constant is a fundamental constant is a lie. The aether deniers, on the other hand treat it as a fundamental constant, since they left themselves no other choice.

The magnetic constant, or vacuum permeability is related to how aether forms magnetic fields due to electric currents or moving electric charges. It is also a property of quantum fluctuations and their ability to form magnetic fields. The magnetic constant is a derivable constant, rather than a fundamental one. **To state that the magnetic constant is a fundamental constant is a lie.** Once again, aether deniers painted themselves into a corner.

The electric constant relates to the polarizability of the aether, while the magnetic constant relates to the magnetizability of the aether. Both require rotation so they are both dependant on rotation within the zero-point field. Van der Waals forces between quantum fluctuation dipoles determine these properties, in particular, the van der Waals torque. Van der Waals torque provides resistance to both linear and rotational motion. And guess what, the speed of light is also derived due to aether's resistance to linear and rotational motion.

So, not only can the speed of light be derived from the electric and magnetic constants but also, all three so-called fundamental constants in equation 87-1 represent different measures of the van der Waals torque of the quantum field.

Returning to Dirac's statement, it turns out that the fine structure constant is not exactly the inverse of 137, so some of the mysticism has gone away. In a set of natural units where h and c are set to 1, the fine structure constant $\alpha = e^2/2$. Since e is the quantum field polarization on a surface around a polarizing particle, α is the polarization of the volume of space around the polarizer.[118] **Neither *e* nor α are fundamental constants.**

Of the four constants Dirac discusses in the quote; Planck's constant is different as it is a scaling factor that changes with the units of energy. Therefore, it is possible to set **h = 1** in natural units and skip it entirely. And since all the other constants Dirac refers to in the equation are a function of the polarizability and magnetizability of the quantum field and the quantum van der Waals torque, so is Planck's constant. It turns out that all physical constants are functions of the polarizability and magnetizability of the quantum field and the quantum van der Waals torque.[119]

The statement that the speed of light is fundamental is a lie. Once we accept that aether is a transmission medium for light, the speed of light becomes a property of that medium, the aether. It is only a matter time before someone derives the speed of light from first principles based on van der Waals forces between quantum fluctuations.

---

[117] P.A.M. Dirac, "The Evolution of the Physicist's Picture of Nature." Scientific American, May 1963, 208(5), 45-53.

[118] R. Fleming, "Fine Structure Constant as the Polarization of the Quantum Field by a Unit Charge." Gsjournal.net. 3 Jul 2018.

[119] R. Fleming, "Physical Constants as Properties of the van der Waals Torque of the Quantum Field." Gsjournal.net. 3 Jul 2018.

Lie #88: Faster than Light Travel

The 88th of the greatest lies in physics is faster than light travel. Faster than light travel has long been a staple of science fiction. Without it, interstellar space travel is impractical. There are numerous physicists who indulge the fantasies of science fiction fans by speculating about ways that the speed of light limit for bodies of matter might be overcome. And, they and their media organizations profit handsomely from it. The allowance of speed of light violations in the big bang model has fueled these fantasies further.

As discussed in previous chapters, the speed of propagation of forces must be much larger than the speed of light, or orbits would be unstable. Evidence for so-called spooky action at a distance also requires that forces and fields propagate much faster than light.

There is even the idea that light can move faster than light if it is transmitted through an elongated Casimir cavity. This is called the Scharnhorst effect, which has not been verified as of yet.[120] The Scharnhorst effect, like the Casimir effect is likely not detectable with shorter wavelengths of light in cavities that are greater than a micron apart. But possibly, in the case of larger wavelengths, like microwaves, the cavity may only need to be slightly larger than the microwave's wavelength. This may be what is occurring in the EmDrive. However, if or when the Scharnhorst effect is experimentally verified, it will still not allow large bodies of matter to violate the speed of light limit.

Some science fiction writers and physicists hypothesize that we could disrupt the zero-point energy in front of a body, thus allowing it to go faster. Others have hypothesized that we could encapsulate the entire ship in order to isolate it from the aether of space, and thus allow it to go faster than light.

The physical problem with these fantastical ideas comes down to the origin of inertia. Inertia is due to the rotation of quantum fluctuation dipoles as a body moves through space. These dipoles can be very small such that they fill the space between the electrons and the atomic nuclei. Some are so small that they fill the space between protons and neutrons in the atomic nuclei, and even the space within the protons, neutrons, and electrons. There is no escaping the ability that quantum fluctuations have to limit the velocity of a moving body.

We also have to consider that the zero-point energy in one cubic centimeter of vacuum is at least equivalent to 10^{95} grams of mass. The energy in a single cubic centimeter is far greater than the estimated non-vacuum energy of the visible universe. If a science fiction writer thinks they have a system that is energetic enough to manipulate the zero-point field, they are sadly mistaken. To be even possible, we would first need to be able to extract large amounts of zero-point energy directly from the quantum field. And large means many orders of magnitude greater than all the mass in the visible universe.

Perhaps the saddest science fiction writers are the physicists who think we can somehow bend space itself. Space is far too energetic for us to bend it, even if it were physically possible to do so, which it is not. There will be more on that problem in a few chapters. **And yes, warp drives are a lie too.**

Faster than light travel is a lie. It is simply not possible to isolate matter from the zero-point field or trick the zero-point field into changing its rest frame. Physicists do the public and science a disservice when they cater to this science fiction idea.

[120] G. Barton, K. Scharnhorst (1993). "QED between parallel mirrors: light signals faster than c, or amplified by the vacuum." Journal of Physics A. 26 (8): 2037.

Lie #89: Tachyons

The 89th of the greatest lies in physics are tachyons. Tachyons are hypothetical particles that move faster than the speed of light. In particular, their slowest possible velocity is the speed of light. Tachyons are a mathematical contrivance that exists outside the normal physical boundaries of the mathematical equations. The reason for tachyons is, surprise, tied to the rejection of aether and the failure to understand, or even attempt to understand the physical mechanics of inertia; that, and mathematicians run amuck.

To be fair, most real physicists do not believe that tachyons really exist. Tachyons are just something else, which exists on the science fiction side of modern physics, and of course, in actual science fiction. This is one idea that deserves to stay firmly in the realm of science fiction.

Tachyons are thought to be the quanta of an imaginary mass field. There is even interest in trying to show that such a field exists. Since mainstream physicists have decided to ignore mass models where mass is a property of zero-point energy, they are still flailing away at it. **Imaginary mass fields are simply a lie within the tachyon lie.**

Tachyons are a lie. Physics would be better off if these hypothetical ideas, including the imaginary mass field, were dropped.

Lie #90: The Remaining Hypothetical Particles

The 90th of the greatest lies in physics is the remaining hypothetical particles. Rather than go though all the hypothetical particles one by one, it is far simpler to dismiss all of them at once. There are dozens of hypothetical particles related to a variety of theories. They sadly have one thing in common. They overcomplicate a particle system, which is already too overly complicated to be fundamentally real

There is a broad group of these particles that is part of the supersymmetry hypothesis based on the electron-neutrino relationship. So physicists came up with photinos, neutralinos, gluinos, gravitinos, higgsinos, and numerous others. **Supersymmetry in general is lie.** There is also the dark photon, a hypothetical carrier of the dark force. There are axions, WIMPS (weakly interacting massive particles), and other particles that are needed to fix problems with theories when those theories are entirely wrong to begin with.

This is not to say that we are done looking for resonances. The onium theory leads to predictions of many more resonances, particularly when we start looking at relativistic mesons and protons. The resonant state of a proton-antiproton pair that is equivalent to a neutral pion will have a mass close to 247 GeV/c^2. There should be numerous other resonant states as well, analogous to the mesons, composed of combinations of proton-antiproton pairs, electron-positron pairs, along with individual electrons, positrons, protons, and antiprotons.

As pointed out in Lie #79, a fundamental theory of physics only needs two fundamental particles, the electron and proton, plus their antimatter opposites. The remaining detectable particles are resonances, and the remaining hypothetical particles are lies.

Lie #91: G is a Fundamental Constant

> *Dilbert:* And we know mass creates gravity because more dense planets have more gravity.
> *Dogbert:* How do we know which planets are more dense?
> *Dilbert:* They have more gravity.
> *Dogbert:* That's circular reasoning.
> *Dilbert:* I prefer to think of it as having no loose ends.[121]
>
> Scott Adams, 1999

The 91st of the greatest lies in physics is G is a fundamental constant. G is the gravitational constant from Newtonian gravity as shown in equation 91-1. We can recall from Lie #26 that gravity is not a fundamental force because it is a composite force with two or three components starting with high-energy gravity pushing objects together and the dark force pushing them apart.

Equation 91-1

$$F = G\frac{m_1 m_2}{r^2}$$

Additionally, there are mechanical Lorentz type forces on bodies with a tangential velocity pushing them toward the center of mass they are orbiting. Bodies moving on parallel paths are also pushed together, just as two parallel wires with parallel currents are pushed together, but due to mechanical forces. A body's rotation introduces yet another component of the mechanical force, which leads to many other phenomena.

Newton's gravitational constant G is similarly the product of two or three constants although it appears that all of the forces, aside from high-energy gravity, can be combined into a single mechanical force. Even high-energy gravity can be considered a component of a combined electromagnetic and mechanical force.

It is clear that a constant derived from two or more different constants is not a fundamental constant. It is possible for the gravitational constant G to be derived from more fundamental force principles. When we factor in that the orbital motion of planets changes the real magnitude of G by way of the mechanical Lorentz force, it is not truly a constant.

You may be thinking that it certainly looks like it is constant when we do the orbital math for each planet. But that ignores the fact that mass is a free parameter since we cannot measure planetary mass independently from gravitational calculations. We simply adjust the mass value of each planet to keep G constant.

The Lorentz force is far more noticeable in spiral galaxies, where there is not enough mass in the galaxy to explain the spirals using Newtonian gravitation alone. More mass, however, would still not explain the spirals.

With stars orbiting the galactic center at more than a hundred thousand kilometers per second, there is a very strong mechanical or non-electric Lorentz type force pushing them inward toward the galactic core. Since two or more stars on parallel paths are also pushed together by the mechanical force, they also form bands. Without these additional forces the outermost stars in a spiral galaxy would fly off into intergalactic space. As a composite 'constant' G takes on a different and much larger magnitude when large mechanical Lorentz forces are added to it. **It turns out that even calling G a constant is a lie.**

The statement that G is a fundamental constant is a lie. G is not even constant. Physicists manipulate the masses of objects and introduce fictional dark matter in an attempt to make G constant, but those fixes do not work. Additional forces are required to explain the dark force and the structure of spiral galaxies.

[121] S. Adams, Dilbert cartoon strip, 1 Mar 1999.

Lie #92: Inverse Square Law for Gravity

The 92nd of the greatest lies in physics is the inverse square law for gravity. Looking back at the last chapter we see an **r²** in the denominator of equation 91-1. This means that the gravitational force is supposed to decrease in proportion to the radius squared, which is more commonly stated as the distance squared. This is the inverse square law.

The inverse square law is also a statement of the principle of conservation of energy. If we consider the energy over the entire surface of several spheres at different distances from the point of origin of a force, the energy of each sphere is identical. This means that the energy is not lost. If energy were lost, without being absorbed by a real physical mechanism, there would be a violation of the principle of conservation of energy.

It is easy to see how this gets us into trouble. The inverse square law works OK within our solar system, which is of course, the evidence Newton used to come up with his theory of gravitation. It does not work so well when we get to spiral galaxies. As described in the last chapter there is not enough mass for Newton's gravitational theory to explain the attraction of the outermost stars. There is also the pesky problem that it does not explain the spiral bands.

Things get stranger when we look at separate galaxies. Instead of all galaxies being pushed together, as we would expect from gravity alone, most, but not all, are moving apart. Gravity fails to hold things together when we get to intergalactic distances. To combat this problem, physicists came up with the idea that there was an explosion that ultimately gives the galaxies their velocity and causing expansion. Since the dark force explains the acceleration, it instead looks like high-energy gravity gets weaker with distance or the dark force gets stronger.

The tiebreaker in this case is that the rate of expansion is accelerating, thus we know that the dark force is stronger than high-energy gravity at intergalactic distances. It also tells us that all of the expansion is due to the dark force, not some residual velocity from some long-ago explosive event.

That brings us back to two options, either gravity gets weaker with distance or the dark force gets stronger, and since the dark force getting stronger would violate the principle of conservation of energy, gravity must get weaker through some mechanism.

Gravity's weakness is that it is transmitted by van der Waals forces, dipoles interacting with other dipoles to produce a pressure force. Gravity is due to bodies of matter interfering with the transmission of these van der Waals forces, creating a shadowing effect reducing the quantum field pressure between the bodies. Matter, however, does not just exist in large bodies, but is spread throughout the universe in otherwise empty space. There are intergalactic particles, gas, plasma, and dust everywhere. That means van der Waals forces are disrupted along their entire path from every direction around each body. Eventually the van der Waals forces become so weak, from all sides, so it does not matter that there is a star or galaxy a long distance away. This is probably why the estimated 10^{53} kilograms of mass in the visible universe do not form a black hole.

The dark force, on the other hand, is consistent with a mechanical force theory. It is the repulsive force between bodies with the same type of charge. The dark force is transmitted due to polarization of quantum dipoles, a mechanism that is not subject to interference from intergalactic gas and dust.

The inverse square law for gravity is a lie. Based on the scientific evidence, gravity declines more rapidly than the square of the distance at intergalactic distances.

Lie #93: Degeneracy Pressure

The 93rd of the greatest lies in physics is degeneracy pressure. Degeneracy pressure is thought to be a type of pressure force that develops between particles, specifically Fermions, due to the Pauli exclusion principle. Within the scope of the Pauli exclusion principle, two Fermion particles cannot have the same quantum state at the same time.

This rule-based system is then supposed to produce a repulsive pressure force, as particles with the same quantum state somehow push each other in order to keep from violating the rule. How particles 'know' not to violate the rule is unclear. This is a supposed force that is not in the standard model list of fundamental forces.

As has been discussed in previous chapters, there are a number of short-range repulsive forces that are not properly accounted for in the standard model of physics. There are repulsive forces between:

1. electrons and protons,
2. electrons,
3. protons,
4. protons and neutrons,
5. neutrons.

From the above list we can see that each of the three particles that make up the stable matter in the universe is repelled from all three types of particles with the possible exception of the neutron and electron. That said, if it has not already been detected, I predict that there is a repulsive force at close range between electrons and neutrons that is similar to the force between protons and electrons.

Getting back to the degeneracy pressure concept, the first degeneracy force is the electron degeneracy force which is said to cause a pressure force between electrons. In the mid to late 1920s physicists were attempting to describe the physics of white dwarf stars. To this end they were modeling them as a gas based on Fermi-Dirac statistics. This Fermi gas accounted for the density of white dwarfs fairly well.

In 1931 Subrahmanyan Chandrasekhar applied this principle and was able to set an upper limit for the mass of a white dwarf star, known as the Chandrasekhar limit.[122,123] Chandrasekhar later won a Nobel Prize for his work along with William Fowler. Within the scope of his hypothesis, above the Chandrasekhar limit, gravity is strong enough to overcome the electron degeneracy pressure and the white dwarf collapses into a neutron star.

If we look back at our list of repulsive forces it is easy to see which one sets the limit for neutron formation. It is the ~780 keV repulsive force between protons and electrons. Once gravity becomes strong enough to overcome this repulsive force, or at least get close enough that weak interactions produce neutrons at a much greater rate than normal, a white dwarf will collapse into a neutron star. **The Chandrasekhar limit is a lie**, at least with respect to its hypothesis. The actual limit is real, but it is due to the repulsive force between electrons and protons.

On a more general topic Freeman Dyson is credited with determining that the solidity of matter is due to electron degeneracy pressure.[124,125,126] It has long been asked why matter feels solid when it is mostly empty space. With physicists ignoring the various repulsive forces between particles, they were generally left with ascribing the solidity of matter to electromagnetic forces. Dyson showed that it makes more sense as an electron degeneracy pressure phenomenon.

In a way Dyson is correct, in that matter feels solid because of a repulsive force between electrons. His mistake is that this force is not due to some pressure force that arises as a result of the Pauli exclusion principle. Electrons do not 'know' the rule; there is a real force responsible for it. Matter feels solid due to the repulsive force between electrons, and the repulsive force between electrons is due to a fundamental force. **Dyson's hypothesis for the solidity of matter is a lie.**

In order to come up with the correct value for the Chandrasekhar limit, physicists had to add a correction to include proton degeneracy, as under the Pauli exclusion principle protons are thought to be subject to a similar pressure force with other protons. This concept of proton degeneracy has been hypothesized as the cause for the repulsive component of the strong force.

In reality protons do not actually 'know' there is a rule either. There certainly is a repulsive force between protons, but it has nothing to do with Pauli's exclusion principle, as it is a true force. We know this because there is a repulsive force between protons and neutrons which is not predicted by the Pauli exclusion principle, as those are two different types of particles.

Physicists then recognized that obviously, there must be neutron degeneracy as well. Accordingly, to them there must be a pressure force between neutrons caused by Pauli's exclusion principle, and it is this hypothetical force that keeps neutron stars from collapsing. As with proton degeneracy, the repulsive force between neutrons is a real force, and not some artifact of Pauli's exclusion principle. Neutrons do not 'know' the rule either.

Rule-based principles are dangerous in physics, particularly when they are based on partial information. The Pauli exclusion principle has already been shown to be a lie in Lie #52, with respect to how it is used to introduce new types of quantum numbers with no basis in physical reality. Here the Pauli exclusion principle is used to explain forces when it is not a force. **Electron, proton and neutron degeneracy are all lies.**

There are related terms that are also lies, as matter under degeneracy pressure is given the name **degenerate matter**. Subsequently there is **electron degenerate matter**, such as the matter in white dwarf stars, and **neutron degenerate matter** in neutron stars. There is also the purely hypothetic **quark degenerate matter** that they think could exist in black holes, but cannot really exist since quarks are not real.

Degeneracy pressure is a lie. There is a real repulsive force between matter particles which is not due to the Pauli exclusion principle. Physicists need to determine how this force arises.

[122] S. Chandrasekhar, "The Density of White Dwarf Stars." Philosophical Magazine, (7th series) 11, pp. 592–596, 1931.
[123] S. Chandrasekhar, "The Maximum Mass of Ideal White Dwarfs." Astrophysical Journal 74, pp. 81–82, 1931.
[124] F. J. Dyson, A. Lenard, "Stability of Matter I." J. Math. Phys. 8 (3): 423–434, 1967.
[125] A. Lenard, F.J. Dyson, "Stability of Matter II." J. Math. Phys. 9 (5): 698–711, 1968.
[126] F. J. Dyson, "Ground State Energy of a Finite System of Charged Particles." J. Math. Phys. 8 (8): 1538–1545, 1967.

Lie #94: Neutron Star Size Limit

The 94th of the greatest lies in physics is the neutron star size limit. The neutron star size limit is usually referred to as the Tolman–Oppenheimer–Volkoff (TOV) limit which states that there is a limit to the size of a body made of neutron degenerate matter. This is analogous to the Chandrasekhar limit for white dwarf stars which was based on the limits for electron degenerate matter.

The TOV limit, as with other degenerate matter, is based on the Pauli exclusion principle model. It was thought that there is a limit to how large a neutron star could be when supported due to degeneracy pressure. J. Robert Oppenheimer and George Volkoff worked out the limit, using methods developed by Richard Chace Tolman.[127,128]

Their original estimate ended up being smaller than the Chandrasekhar limit for white dwarf stars at 0.7 times our sun's mass. The TOV limit was later reworked to include the repulsive force between neutrons, falsely attributed to the strong force, to raise the estimate to around 1.5 times the mass of our sun. That way the TOV limit became greater than the Chandrasekhar limit of ~1.4. Other, more recent estimates range up to 3 times the sun's mass. Neutron stars have been observed in the 1.4 to 2 solar mass range, overlapping somewhat with white dwarfs on the low side.

In case you missed it, in order to get a correct value, they had to resort to using the repulsive force between neutrons, rather than degeneracy pressure. Ultimately it is the repulsive force between neutrons, which leads to the correct mass for neutron stars, not degeneracy pressure. Even if degeneracy pressure were a real thing, the important force determining the size of neutron stars would be the repulsive force between neutrons, a force that is not included in the standard model.

And guess what, there is no known limit on the neutron repulsive force. As far as we know, both from our models and experiments, there is no point where the repulsive force between neutrons, or the same force between protons, or protons and neutrons, fails.

The experimental evidence does not show that protons can be superimposed on top of other protons. The evidence does not show that protons can be superimposed on neutrons or vice versa. The evidence does not show that neutrons can be superimposed on other neutrons. Atomic nuclei do not magically collapse on random occasions, as we might expect from quantum theory, if it were possible.

Somehow physicists ignore the facts in evidence, because they want to believe that particles could be further compressed. We live in a quantum world and if such superimposed states existed, we would have seen them, if only infrequently. You can think of it being statistically similar to the decay of a proton. If it occurred ever in the history of the universe, we would have seen it already in the laboratory.

There is yet another way to look at the problem. We can consider the gravitational force inside a sphere, as this tells us about the gravitational forces on matter inside a black hole immediately after it forms. What is important to know is that the gravitational force inside a spherical shell of matter and energy, due to the shell, is zero everywhere inside the shell.

That means that matter at a larger radius does not gravitationally affect matter inside that radius. Consequently, once a black hole forms there is no new gravitational force exerted on the material inside. There is no new matter or energy weighing on the neutron star once it becomes a black hole, as all matter and energy becomes stuck at where the event horizon was when the material entered the black hole.

Neutron stars do not collapse into another form of more dense material. There is no form of more dense material to collapse into. There is no extra gravitational force on the neutrons. There is no additional pressure on a neutron star inside a black hole. There is no known quantum mechanism for overcoming the repulsive force between neutrons, even if there was a force to cause it.

Neutron stars simply continue to grow larger until they become a black hole, and then the core of the black hole only grows as matter and energy accumulate where the event horizon was at their time of absorption.

The neutron star size limit (TOV limit) is a lie. There is no known force that limits the mass of neutron stars, and no evidence that two neutrons can exist in a superimposed state.

[127] R.C. Tolman, "Static Solutions of Einstein's Field Equations for Spheres of Fluid." Physical Review, 55 (4): 364–373, 1939.
[128] J.R. Oppenheimer, G.M. Volkoff, "On Massive Neutron Cores." Physical Review, 55 (4): 374–381, 1939.

Lie #95: Relativistic Black Holes

The 95th of the greatest lies in physics is relativistic black holes. Black hole physics gets lots of attention in popular science and science fiction. Artists have designed tantalizing images to awaken people's imaginations and cause wonderment. Physicist-entertainers and people who hire them make a bundle of money off black holes.

John Michell, first discovered black holes, or as he called them black stars, in 1783.[129] He was able to predict black stars from a purely classical perspective, based on the corpuscular theory of light. The idea that light traveled in discreet packets, or quanta, was not really new in 1905, just improved due to Planck's theory. Michell was able to predict that there are numerous black stars.

In the 1900s, Karl Schwarzschild was the first to find an exact solution to Einstein's equations of general relativity in 1915 where he described the physics that later led to relativistic black hole theory.[130] Although, it was John Archibald Wheeler who coined the term 'black hole' much later.

Working from Michell's ideas and Schwarzschild's equations, we find that there can be an ultra-dense body where nothing, not even light, can escape its gravitational force, at least not in a classical way. The equation for the radius of such a body, called the Schwarzschild radius, is shown in equation 95-1.

Equation 95-1

$$r_s = \frac{2GM}{c^2}$$

It is important to note that, as John Michell showed, this equation can be derived in a purely classical way without resorting to the hocus pocus of Einstein's general relativity theory.

Also note the 'constant' G, which is not really a constant, and the not so fundamental constant, the speed of light, are in the equation. Both 'constants' G and c are more fundamentally described as properties of quantum fluctuations. This means that we need to reexamine them in the context of black holes, but I will not do that here.

While the gravitational force around such an ultra-dense object is unquestionably large, there is no warped space-time. **Warped space-time is another lie.** The space in and around a black hole is still geometrically flat, so much for all that fancy artwork.

The ultra-dense precursors of black holes are neutron stars, which were discussed in the last chapter. Neutron stars form when the gravitational energy is sufficient to overcome the force between protons and electrons. These neutrons are then pushed together by the strong nuclear force, a strong Casimir Force, until the force that opposes the strong force starts at a distance of about 0.7×10^{-15} meters between the neutrons.

Physicists hypothesize that neutron stars gravitationally collapse when they reach between two and three solar masses. For the reasons outlined in the last chapter the collapse of large neutron stars is a lie. When they are large enough, neutron stars become classical black holes. The gravitational collapse of relativistic black holes is unproven and unnecessary. And, if the entire visible universe is inside a black hole as the basic theory suggests, we know that the relativistic theory of the interior of a black hole is wrong.

It is also important to note that it would take forever for a photon to leave the event horizon. And, it would be nearly infinitely redshifted and undetectable. A classical black hole would still be black, as no light can escape the event horizon.

The speed of light limit means that gravity, within the standard model, would cease to exist at the event horizon, as gravitons, or whatever, could not exceed the speed of light. We can consider it a good thing that gravity propagates at faster than the speed of light. If not, gravity would not be detectable outside a black hole's event horizon. Fortunately, quantum fluctuations only need to rotate a small amount for forces to be transmitted, even under such extreme conditions.

Gravitational mass is like relativistic mass, not limited to the rest mass. So the effective mass goes to infinity at the event horizon, as the clock rates go to zero. So once a classical black hole forms it starts to develop a shell of high-density energy in its growing event horizon. A classical black hole becomes a black sphere of sorts.

The relativistic black hole hypothesis is another of the non-scientific religious hypotheses that have infected physics as:

A. There is no physical evidence for them.
B. The relativistic model is unnecessary to describe black holes.
C. The relativistic model adds unnecessary complications to black hole theory.
D. The collapsed interior of a relativistic black hole can never be detected.

The relativistic black hole model is a lie. It is time we took a different approach to black hole science and stop disseminating the sensationalistic science fiction propaganda.

[129] J. Michell, "On the means of discovering the distance, magnitude etc. of the fixed stars ...", Philosophical Transactions of the Royal Society of London, The Royal Society, 74: 35–57 & Tab III, 1784.
[130] K. Schwarzschild, "Über das Gravitationsfeld eines Massenpunktes nach der Einsteinschen Theorie." Sitzungsberichte der Königlich Preussischen Akademie der Wissenschaften. 7: 189–19, 1916.

Lie #96: Hawking Radiation

The 96th of the greatest lies in physics is Hawking radiation. Stephen Hawking proposed this black hole radiation hypothesis, which bears his name.[131] The basic concept requires quantum fluctuations, which was surprising at the time, since aether deniers dominated the field of physics. For some unknown reason they allowed it in this case, while still ignoring it in most other fields of physics.

The basic idea is that when a quantum fluctuation particle pair comes into existence just outside the event horizon of a black hole, in some cases one of the particles of a particle pair crosses the event horizon. Once the quantum fluctuation crosses the event horizon it cannot cross back. Hawking's idea was that in some cases the other particle would remain free outside the event horizon, and since he considered these pairs to be photons, this free photon had a chance to escape the gravitational pull of the black hole while traveling at the speed of light.

He thought an escaping virtual photon would carry away energy, and since he thought the energy had to come from somewhere other than the zero-point field, the energy had to come from the black hole. In this way the black hole could radiate energy. Over the years, as quantum effects at the event horizon of a black hole have been studied further, there are questions of whether a photon could actually escape. Black holes cannot actually radiate.

The problem with the Hawking radiation hypothesis is worse than that. As discussed back in lies #5 and #6, virtual photons do not exist, and consequently the aether is not made of virtual photons. Instead, aether is made of electron-positron and proton-antiproton particle pairs. Reconsidering the Hawking radiation hypothesis with particles that have charge and mass when free and stable changes things. Massive particles do not travel at the speed of light, but at best slightly less than it.

Consequently, the whole idea of newly freed electron or proton escaping the gravitational pull of a black hole from just outside the event horizon is nonsense. If a quantum electron-positron pair forms just outside the event horizon and one falls in, it is only a brief matter of time before the other one falls in as well. Of course, another problem is that the amount of time it takes a particle to fall in is forever. A particle only crosses the event horizon when the event horizon grows past it.

Ignoring the forever problem for a moment, the two particles can cross-annihilate with another particle pair. We can imagine in rare cases, two electron-positron pairs, side by side, such that one pair's electron and the other pair's positron meet at the event horizon and annihilate. This leaves an electron and positron slightly outside the event horizon. If that pair annihilates before exceeding the Planck and Heisenberg limits, then their energy returns to the vacuum.

If they somehow exist for too long, they become free particles with mass, and take energy from the black hole or the zero-point field. We do not really know which. Then if they somehow manage to annihilate with each other before being stuck at the event horizon, they would produce a photon. That photon in turn has a chance to escape, but it would take a nearly infinite amount of time and it would be nearly infinitely redshifted down to essentially zero energy, so in essence no energy could escape anyway.

Hawking radiation is a lie. Black holes do not radiate energy by absorbing half of a virtual photon pair.

[131] S. Hawking, "Black hole explosions?" Nature 248, 30, 1974.

Lie #97: Singularities

The 97th of the greatest lies in physics is singularities. Singularities are thought to occur within the standard model when a body collapses when its mass becomes extremely large. Physicists believe that once the limit of neutron degeneracy pressure (Lie #94) is reached all matter will collapse, first to degenerate quark matter and then to a singularity. At that point essentially all of the energy is compressed to a dimensionless point. These are sometimes referred to as gravitational singularities or space-time singularities.

The most commonly discussed singularities are in the center of a relativistic black hole (Lie #95) or at the beginning of a big bang (Lie #34). But singularities are said to occur in any body that can overcome all forms of degeneracy pressure (Lie #93). Singularities often occur as an artifact of curved space theorems (Lie #43), particularly within the scope of general relativity theory (Lie #44).

A pattern appears to be emerging. Singularities are sort of like paradoxes. If a model leads to paradoxes then the model is a lie. **If a model has singularities then it is a lie.**

Singularities are lies. The repulsive force between particles prevents matter from collapsing. Singularities are science fiction.

Lie #98: Wormholes

The 98th of the greatest lies in physics is wormholes. Wormholes, more technically known as Einstein-Rosen bridges, are a hypothetical feature due to general relativity (Lie #44) that are thought to link two points in space-time. Wormholes are thought to allow faster than light travel (Lie #88) across vast distances in space, and as such they are very popular among science fiction fans, and the people who make money off them.

The wormhole idea begins with a relativistic black hole (Lie #95), which allows matter in but not out. By extending the black hole equations one can imagine a hypothetical white hole that only lets matter out. This is a case of the mathematicians are doing their science fiction thing again.

Then if somehow a black hole and this mythical white hole are linked through space-time, it is hypothesized, or more properly fantasized, that someone could travel in one end and out the other, as long as they do not run into a singularity (Lie #97). The white hole would eject them into another point in space and time. Mathematical physicists have then toyed with the math to find ways to make them stable and to allow two-way travel.

While wormholes may be fun to fantasize about, they exist firmly in fantasyland and science fiction. General relativity, on which the idea is based, is a lie, relativistic black holes are a lie, singularities are a lie, and yes, **white holes are a lie too.**

Wormholes are a lie. They are nothing but a mathematical artifact of a bad theory. They should remain in the realm of science fiction and not to be confused with real physics.

Lie #99: Matter Production Violates the Principle of Conservation of Energy

The 99th of the greatest lies in physics is matter production violates the principle of conservation of energy. Perhaps the greatest remaining questions in physics relate to the production of matter.

   a. Where does matter come from?
   b. What is the source of the energy?
   c. How is matter produced?
   d. How is there only matter and not antimatter?

The favorite mainstream answer to all these questions is the big bang model (Lie #34), but big bang models do not actually solve any of these problems.

As for the principle of conservation of energy, it states that energy cannot be created or destroyed. Energy can, however, be converted from one form to another, including mass. The use of fission, fusion, batteries, and hydrocarbon fuels are all examples of converting one form of energy to another. Energy, including matter, cannot be produced from nothing.

Once again, we have a problem created by the aether deniers. They thought that space was empty, a complete vacuum, with no aether. So of course, matter cannot come from such an empty space. But they take things one step further and invent the idea of non-space (Lie #28) that a big bang can grow into, and somehow that would be OK. It is not OK since non-space isn't real. **Matter produced from non-space is a lie.** That truly would be a violation of the principle of conservation of energy.

We can say that since matter exists, we know matter is produced, somehow. It did not appear by magic. And, space is filled with zero-point energy, much more than necessary to account for all the mass-energy in the visible universe. But here the aether deniers say that zero-point energy does not count. They do not include zero-point energy in their thinking about the principle of conservation of energy. To them, energy from zero-point energy is also a violation of the principle.

To say that energy is not energy with respect to the principle of conservation of energy is another lie. Energy is energy, and one form may be converted to another if there is a physical mechanism that allows for it to happen. **Some physicists have been known to call zero-point energy 'negative energy,' but that is still another lie.** Like all forms of energy, zero-point energy is positive energy. **Negative energy is another lie by itself.**

It turns out that it is easy to see how electrons, protons, and neutrons can come from zero-point energy. The vacuum is filled with quantum electron-positron and proton-antiproton dipoles, so we know that the particles are already there. It is strictly a matter of getting the matter particles out, separately from the antimatter particles. Once you have electrons and protons, they will combine to form neutrons.

We have also found that the masses of the proton, neutron, and electron are equivalent to the zero-point energy they displace. The amount of energy in a volume of space is unchanged whether particles are present or not. We can even amend our statement of the principle of conservation of energy to state that the total energy in a region of space is constant. This is true for all forms of energy; however, there is not space to go through the details here. Please refer to my paper, "a zero-point energy inclusive principle of conservation of energy consistent with zero net energy in the universe."[132]

Not only are the particles already present, but when they become free particles, they do not affect the total energy of space. The production of protons, neutrons, and electrons from zero-point energy does not violate the principle of conservation of energy. **The statement that**

matter cannot be produced from zero-point energy is a lie.

That answers questions a and b, which leaves us with c and d. Unfortunately, I do not have an answer for you today, and while I could speculate wildly, I will resist the temptation.

What we know is there is some way to extract electrons and protons from the quantum field. Figuring out how is perhaps the greatest single problem in physics and yet hardly anyone is working on it. At least hardly any experimentalists are working on it, and that is where the breakthrough must come. Nobody will believe a matter production hypothesis until somebody can demonstrate matter production in the laboratory. This is an area of research that deserves more attention and funding. We should be able to figure out a way to produce matter for a fraction of what it costs to produce Higgs resonances.

The statement that matter production violates the principle of conservation of energy is a lie. We know that matter must be produced. It is only a matter of time before the mystery of how matter is produced is solved.

[132] R. Fleming, "A zero-point energy inclusive principle of conservation of energy consistent with zero net energy in the universe." Researchgate.net.

Lie #100: There are Four Fundamental Forces

> *Behind it all is surely an idea so simple, so beautiful, that when we grasp it – in a decade, a century, or a millennium – we will all say to each other, how could it have been otherwise?*[133]
>
> John Archibald Wheeler 1986

The 100th of the greatest lies in physics is there are four fundamental forces. In the standard model of physics there are said to be four fundamental forces:

    a. electromagnetic force,
    b. gravitational force,
    c. weak force,
    d. strong force.

Physicists have attempted to describe all force interactions using just these four forces, while at the same time trying to unify them. The electric and weak force are already said to be unified based on the weak interaction model where those interactions are moderated by the W and Z bosons. So the standard model actually considers there to be three forces left to unify.

As we went through the various lies, we made a more comprehensive list of forces. They are the:

    1. mechanical force,
    2. electromagnetic force,
    3. weak force,
    4. strong force,
    5. dark force,
    6. high-energy gravity force,
    7. repulsive force between electrons and protons,
    8. repulsive force between electrons,
    9. repulsive force between protons,
    10. repulsive force between protons and neutrons,
    11. repulsive force between neutrons,
    12. attractive force between electrons.

To that list we could add degeneracy pressure (Lie #93), the Pauli exclusion principle force, except it is redundant with respect to numbers 8, 9, and 11, the repulsive forces between electrons, protons, and neutrons. We can also hypothesize that there are close range repulsive forces between electrons and neutrons, given that all the other particle combinations of the three primary particles already are known to exhibit a repulsive force.

Starting with the mechanical forces, it is clear that they are true forces, not pseudo-forces, and must be on the final list in some fashion. We will return to them later. The electromagnetic forces appear to be the most fundamental of all the forces and must definitely stay on the list. Because charged objects have mass and matter a complete electromagnetic theory includes the non-electric mechanical forces.

The gravitational force is not a fundamental force as it is the summation of three forces, the mechanical force, the dark force, and the high-energy gravity force. It was therefore left off the second list.

The high-energy gravity force appears to be a Fatio-Casimir force. The existence of the Casimir force tells us quantum fluctuations exist and they push on matter. It also tells us that matter behaves inelastically to this pressure force, as required by Fatio's theory. Additionally, the Casimir force tells us that bodies do not heat due to interactions with quantum fluctuations. All the major flaws with Fatio's theory are not flaws at all. The existence of the Casimir effect tells us that the Fatio-Casimir effect must also exist. This means that Fatio-Casimir high-energy gravity is not a separate force, but an electromagnetic and mechanical effect, and part of a combined electro-mechanical force.

The weak force, or more properly the weak interactions are readily explained as an electromagnetic force, however, not in the way described by the standard model. Weak interactions, such as nuclear decay, are readily explained by annihilation and production events between particles and quantum particle pairs. To the extent that this is a force, it is part of quantum electrodynamics within electromagnetic theory. **It is a lie to call the weak force a fundamental force.**

Next are the strong forces. The strong forces are also a combination of two forces, a strong attractive force and a strong repulsive force. The repulsive forces are listed separately in numbers 9, 10 and 11. As for the strong attractive force it is readily shown to be the Casimir force, which is an electromagnetic force, and not a separate force at all. **It is a lie to call the strong force a fundamental force.**

The dark force is the long-range repulsive force between bodies of matter which is responsible for the accelerating expansion of the universe. The mechanical force includes a mechanical dipole, which is almost certainly matter and antimatter. In any case, the mechanical force requires that matter be repelled from matter. Even if the dipole were something different, bodies of matter are still repelled from each other as shown in 7 through 11. The dark force then is nothing more than a component of the mechanical force.

We can lump the repulsive forces between particles 7, 8, 9, 10, and 11 together as they must be related. These forces all appear to be part of a fundamental repulsive force between matter particles. Once again these forces are consistent with the electromechanical force.

Note that the mechanical force also requires that matter be attracted to antimatter, such that electrons are attracted to positrons, protons attracted to antiprotons, etc... Since particles annihilate with their antiparticles, that part of it makes sense, as they are not repelled from each other at short distances. We also see annihilation and particle decay when an antiproton or positron interacts with a neutron, which is indicative of an attractive force rather than a repulsive force.

That leaves us with a close-range attractive force occurring between positrons and protons; however, since they are both electrically positive, the electric charge repulsion dominates the interaction, at least until they get very close. Also, in normal matter, positrons annihilate with electrons long before reaching a proton. Nonetheless, it would be something to look for. Taken all together it is clear to see that the repulsive forces between particles truly must be a part of the electromechanical force. The details of this interaction still need to be worked out.

Last is the strong attractive force between electrons, which is another strong Casimir force and part of electromagnetic theory.

That leaves us with only two forces. But, since electrons and protons have electric charge and mechanical charge, it is not possible to separate the two forces. They are also mathematically equivalent as Maxwell's equations, and a variation on Maxwell's equations describe both forces. Both forces also behave similarly at the quantum level. Due to this, the two forces are readily combined into a single electromechanical force, or as I have called it elsewhere the electro-matter force.[134]

The statement that there are four fundamental forces is a lie. Physicists have long ignored or been mistaken about numerous forces, which have been verified experimentally. Ultimately however, all forces are a part of the extended electromagnetic force.

---

[133] John Archibald Wheeler, Annals of the New York Academy of Sciences, 480 (1986)
[134] R. Fleming, "The Electro-Matter Force," researchgate.net.

What Now?

> *Science alone of all the subjects contains within itself the lesson of the danger of belief in the infallibility of the greatest teachers in the preceding generation . . . Learn from science that you must doubt the experts. As a matter of fact, I can also define science another way: Science is the belief in the ignorance of experts.*[135]
>
> Richard Feynman

Feynman certainly got that right, but even he would not have anticipated just how bad modern physics has become. And, this list did not include many other theories currently under critical review. Additionally, under each general physics topic, there are dozens of subtopics that are invalid science, and only a few of those have been touched on here. The vast majority of physics papers being published today are a waste of time and effort as they are destined to be relegated to the dustbin of history, for many papers that is the day they are published.

We must recognize that the system behind the science of physics is broken. There is an unnatural divide between the academic community and independent researchers. The academics surround themselves with yes men who do not openly question their authority, even though most are in agreement with Feynman on the necessity of doing just that. Independents are shut out from being published in journals and having access to persons in the academic world for the purpose of serious discourse. Independents are shut out from presenting at mainstream conferences as well.

The journal system is equally broken, with editors refusing to consider ideas that challenge the mainstream view, even when it is known to be wrong. Journals also prevent the sharing of information, by limiting access to articles. People who do not have a major university nearby and cannot afford the large fees cannot get copies of papers. The sharing of articles and information needs to be on an open platform for easy sharing, with a more open system of licensing content. The peer review system is broken anyway; it might as well be an open system instead of one marred with censorship.

As for what is next, the first step is to openly recognize that aether exists and has a dipole character. And then, every theory needs to be reviewed to make certain that it is consistent with the existence of quantum fluctuation dipoles. I tried to do some of that here, and more detailed information can be found in my book *The Zero-Point Universe*.

Then we must face up to the basic unanswered questions, such as how bodies move in response to electromagnetic, mechanical, and gravitational forces. Bodies do not move by magic, there is a real physical mechanism, and that mechanism involves quantum fluctuations.

Then we need to understand quantum problems, such as quantum jumps, low-energy nuclear reactions, photon formation, and the nature of how probabilistic events yield singular results.

There is no magic in physics. The more we insist on understanding the physical mechanisms for everything that happens at the most basic levels, the more we constrain the problems. In the end, we are only left with a single solution. Some of those solutions are reported throughout this book.

And let us get rid of the science fiction in physics, or at the very least, categorize it as such. Discussing science fiction as if it were real science in the popular media prevents real physics research from being taken seriously.

The science fiction problem is also part of something more general. Physicists tend to present theories without going into details about the theory's shortcomings. As the 100 lies show, it is easy to find a fatal flaw with many mainstream physics theories and yet those flaws are seldom discussed openly. Physicists hate to openly admit that favorite theories are garbage, but that is the first step toward moving forward.

We also do not move forward by inventing new things, such as particles, fields, and dimensions that do not exist, in an attempt to correct errors or fill gaps in a theory. We move forward by explaining how things physically work, using what we know to be physically real.

And remember, stop calculating and find the physical explanation.

[135] R. P. Feynman, *The Pleasure of Finding Things Out.* p.186-187, 1999.

Dishonorable Mentions

Less than great lies	From Lie#
Aether heats bodies of matter	1
Aether causes kinetic energy loses	1
There is no aether rest frame	2
Maxwell's aether drag hypothesis	2
Light travels in the reference frame of its source	2
The speed of light is dependant on direction	2
Waves are made of waves	3
Photons are particles	4
Electromagnetic fields are intrinsic to photons	4
Photons are antiparticles	5
Electromagnetic action at a distance	8
Any big bang model that requires magnetic monopoles	10
Any matter production hypothesis that requires magnetic monopoles	10
Any hypothesis that requires magnetic monopoles	10
The speed of light is intrinsic to light	11
Relativistic frame transformations can be performed without a standard reference frame	13
Theories that include unresolved paradoxes	13
Length contraction is needed to explain the Michelson-Morley result	14
Empty space has a clock	16
Proton mass is a fundamental constant	19
Electron mass is a fundamental constant	19
Neutron mass is a fundamental constant	19
All the energy in the universe in one place would not produce a black hole	29
The big bang model explains matter production	32
Non-infinite cosmological models with respect to time do not violate the principle of conservation of energy	33
The big crunch	35
Physics explains photon formation	37
Physics explains how mass causes gravity	52
Point charges	53

Electrons do not have structure	53
Particles in orbits or traveling along a curved path must radiate photons	58
All quark meson models	60
Physicists understand the cause of beta decay	64
Physics explains the cause of weak interaction energy distributions	64
Neutrinos cause the beta decay energy distribution	64
W and Z particles are elementary	68
Mesons mediate the strong force	69
Pauli's exclusion principle is responsible for repulsion between nucleons	69
The repulsion between protons and neutrons is part of the strong force	69
Lepton conservation	72
A particle is not elementary if it has structure	73
Proton decay	73
Relativistic mass	75
The Higgs field	77
Multiverses	82
Curled up extra dimensions	84
Superstrings	85
M-Theory	85
D-Branes	85
Particles can be modeled as strings	85
Time is a physical spatial dimension	86
The electric constant is fundamental	87
The magnetic constant is fundamental	87
e is a fundamental constant	87
α is a fundamental constant	87
Warp drives	88
Imaginary mass fields	89
Supersymmetry	85
G is constant	91
The Chandrasekhar limit	93
Dyson's hypothesis for the solidity of matter	93
Degenerate matter	93
Electron degeneracy	93
Electron degenerate matter	93
Proton degeneracy	93

Neutron degeneracy	93
Neutron degenerate matter	93
Quark degenerate matter	93
Warped space-time	95
Large neutron stars collapse	95
A theory can include singularities	97
White holes	98
Matter can be produced from non-space	99
Zero-point energy is not energy	99
Zero-point energy is negative energy	99
Negative Energy	99
Matter cannot be produced from zero-point energy	99
The weak force is a fundamental force	100
The strong force is a fundamental force	100